Life of the Flycatcher

Animal Natural History Series

Victor H. Hutchison,
General Editor

Life of the Flycatcher

By
Alexander F. Skutch

Illustrated by
Dana Gardner

University of Oklahoma Press : Norman and London

Also by Alexander F. Skutch and illustrated by Dana Gardner

A Bird Watcher's Adventures in Tropical America (Austin, 1997)
A Naturalist on a Tropical Farm (Berkeley, 1980)
New Studies of Tropical American Birds (Cambridge, Mass., 1981)
Birds of Tropical America (Austin, 1983)
Nature through Tropical Windows (Berkeley, 1983)
Life of the Woodpecker (Ithaca, 1985)
*Helpers at Birds' Nests: A Worldwide Survey of Cooperative Breeding
    and Related Behavior* (Iowa City, 1987)
A Naturalist amid Tropical Splendor (Iowa City, 1987)
Life of the Tanager (Ithaca, 1989)
(with F. Gary Stiles) *A Guide to the Birds of Costa Rica* (Ithaca, 1989)
Life of the Pigeon (Ithaca, 1991)
Origins of Nature's Beauty (Austin, 1992)
The Minds of Birds (College Station, 1996)
Orioles, Blackbirds, and Their Kin (Tucson, 1996)
Antbirds and Ovenbirds: Their Lives and Homes (Austin, 1996)

LIBRARY OF CONGRESS CATALOGING-IN-PUBLICATION DATA

Skutch, Alexander Frank, 1904–
    Life of the flycatcher/by Alexander F. Skutch; illustrated by Dana
    Gardner.
        p.   cm.—(Animal natural history series; v. 3)
    Includes bibliographical references and index.
    ISBN 0-8061-2919-0 (alk. paper)
    1. Tyrannidae. I. Title. II. Series.
    QL696.P289S58   1997
    598.8'23—dc21   96-47006
                                                            CIP

Text design by Debora Hackworth.

Life of the Flycatcher is Volume 3 in the Animal Natural History Series.

The paper in this book meets the guidelines for permanence and dura-
bility of the Committee on Production Guidelines for Book Longevity of
the Council on Library Resources, Inc. ♾

To the memory of Laca,
the Gray-headed Chachalaca
who sat at my feet while I wrote
  this book,
and her brother Chacha,
who rested on the windowsill

Contents

Illustrations

PLATES

Following page 50

1 Ochre-bellied Flycatcher
2 Common Tody-Flycatcher
3 Many-colored Rush-Flycatcher
4 Golden-crowned Spadebill
5 Royal Flycatcher
6 Sulphur-rumped Flycatcher
7 Ornate Flycatcher
8 Yellow-bellied Flycatcher
9 Vermilion Flycatcher

DRAWINGS

x   Illustrations

TABLES

Preface

The largest family of birds confined to the Western Hemisphere comprises a fascinating diversity. From whatever aspect we view them—size, coloration, diet, mode of foraging, mating habits, nests—we find great contrasts. In size they range from extremely small to large for passerine birds. Although many are plainly clad, a few are colorful. They eat fruits as well as insects, which they may catch high in the air, glean from foliage, or gather from the ground. Although most breed in monogamous pairs, a few have strange mating habits. Probably no other family of birds (with the possible exception of the ovenbirds) builds more diverse nests. In their wide diversification, species of flycatchers have come to resemble birds of quite different families, making a survey of their habits an instructive overview of bird life.

To distinguish them from the unrelated flycatchers of the Old World in the family Muscicapidae, these New World birds are commonly known as the tyrant flycatchers, or Tyrannidae. This designation is derived from the name *Lanius Tyrannus* that in 1758 Linnaeus gave to the Eastern Kingbird, which he mistakenly classified with the shrikes. Not only is this name unfair to the Eastern Kingbird and its close relatives, but for the family as a whole it is grossly misleading. Although bold in chasing, and frequently buffeting, birds much bigger than themselves that might prey upon their eggs or young, most kingbirds live amicably amid their harmless feathered neighbors. Many flycatchers are among the mildest, least offensive of birds. American or New World flycatchers would be more appropriate names for the family.

Tyrant, or its diminutive *tyrannulet*, occurs in the name of no species in the United States or Canada and of few in Mexico and Central America. Many of the South American birds called tyrants were so designated in *Argentine Ornithology* (1888) by W. H. Hudson. Years later, in *Birds of La Plata* (1920), he expressed the hope that better names could be found for some of the hundreds of species of birds for which he had to invent English names. For uniformity in this book, I have substituted *flycatcher* for *tyrant* or *tyrannulet* wherever these have been applied to South American members of the family (including for such names as marsh-tyrant, tody-tyrant, and pygmy-tyrant).

One of the very first birds I studied after coming to the tropics was a member of this family, the charming little Common, or Black-fronted, Tody-Flycatcher, in a Panamanian garden. In subsequent decades I have watched at the nests of some forty other species of flycatchers, great and small. Because they are so abundant and their nests often accessible and not as difficult to find as those of many other tropical birds, flycatchers have claimed more of my attention than birds of any other family, except possibly hummingbirds and tanagers. In the United States and Canada a number of flycatchers have been carefully studied by amateur and professional ornithogists. In South America, where flycatchers are more abundant in species and individuals, a much smaller proportion of them have been carefully observed. Hence, although we know much about the lives of flycatchers, many of the most eccentric tropical species remain to be studied by naturalists.

The plan of this book follows closely that of its predecessors, *The Life of the Hummingbird, Life of the Woodpecker, Life of the Tanager,* and *Life of the Pigeon.* As in those books, I have placed in the index, for readier reference, the scientific names of all living things for which English names are capitalized in the text.

Life of the
Flycatcher

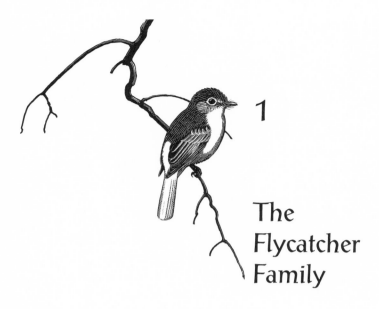

1

The
Flycatcher
Family

Of all my memories of flycatchers, one that remains most vivid is of an April morning I passed years ago watching a Boat-billed Flycatcher incubate in the top of a small silk-cotton tree in a pasture. The tree was renewing its foliage with bright red young leaflets, tinged with bronze and green. The big, heavy-billed bird—quietly sitting on her eggs with her boldly marked black-and-white head raised above the nest's rim and displayed amid brilliant foliage against an azure sky—made such an alluring picture that I longed for the ability to paint her in her red-and-blue setting, with a few white clouds gleaming in the sky above.

Whenever the female Boat-bill left her nest to forage, her mate perched nearby to guard it. With warning clacks of his black bill, he held aloof such harmless intruders as a Vermilion-crowned Flycatcher and a wintering Yellow Warbler. On the female's return, the partners greeted each other with loud, clear notes. Then he flew away while she settled on her eggs with soft twitters expressive of contentment. When a party of big Chestnut-mandibled Toucans appeared upon a hilltop five hundred feet distant, she jumped from her eggs to help him drive away these huge-billed birds, which often plunder nests of Boat-bills and other birds. To repel potential predators before they come dangerously near is the far-sighted policy of Boat-bills and kingbirds. Although these biggest flycatchers in Central America so boldly attack despoilers of their nests, I have never known them to harm any smaller bird.

This experience of long ago and many similar ones point to an outstanding characteristic of the flycatchers that has contributed greatly to their success: vigilant care of their young. Whereas the bigger species practice active defense, the smaller and weaker of them, adopting a different strategy, laboriously construct elaborate nests that are difficult for predators to find or to reach. Yet even some of the smallest flycatchers defiantly menace intruders many times their size. Some avian families are celebrated for the splendor of their plumage, others for their songs. Few flycatchers are brilliantly attired, few sing melodiously; their claim to distinction is the quality of their parental care.

With about 380 species, the American flycatchers, or Tyrannidae, are the largest family of birds confined to the New World, exceeding the next largest family, the hummingbirds or Trochilidae, by about 50 species. From north of the Arctic Circle in Alaska and the Yukon, where Say's Phoebes, Alder Flycatchers, and Olive-sided Flycatchers nest, flycatchers range over the continental Americas to Tierra del Fuego. They are distributed throughout the West Indies and have become established on such distant islands as Cocos, the Galápagos, the Falklands, and Juan Fernández, on each of which at least one species breeds. Like other exclusively New World families, they are most abundant and diverse in South America. Canada and the continental United States together have 35 species, many of them confined to the southern U.S. border regions. Mexico has 66 species, Panama has 89, Colombia 183, Venezuela 154, Peru 224, Brazil 185, and Chile 32. In this vast region, flycatchers are established in almost every habitat: tropical rain forests and dry woodlands, coastal mangroves, temperate deciduous forests, northern coniferous forests, swamps and marshes, savannas, arid deserts, and from warm lowlands to cold treeless páramos and punas up to 15,000 feet (4,600 meters) in the Andes.

Of Earth's approximately nine thousand species of birds in twenty-eight orders, well over half belong to the single order Passeriformes, the passerine or "sparrowlike" birds, a large assemblage of families for which limits are often poorly defined, with the result that their number and arrangement are continually changing as new classifications are proposed. The passerines, in turn, have two major divisions: the oscines or songbirds, which include such familiar birds as swallows, thrushes, finches, and jays, and the suboscines. To this second, much smaller division belong the American flycatchers, in company with woodcreepers, ovenbirds, antbirds, tapaculos, cotingas, tityras, and manakins, all confined to the New World, and a few small families in the Old World. A principal difference between the oscines and suboscines is the simpler musculature of the syrinx, or vocal organ, of the latter, making these birds, on the whole, inferior vocalists, although some of them succeed in producing melodious songs. In nest construction and parental care, the suboscines are not inferior to the oscines; indeed, many of their constructions are more elaborate than those of songbirds.

In size, flycatchers range from the stub-tailed Black-capped Pygmy-Flycatcher, only two and a half inches long and one of the smallest of all birds, to the Great Shrike-Flycatcher of southern South America, one race of which measures a foot in length. If projecting tail feathers are included, the longest flycatchers are some Streamer-tailed Flycatchers, seventeen inches overall.

Black-capped Pygmy-Flycatcher, *Myiornis atricapillus*. Sexes alike. Costa Rica to western Ecuador.

Most flycatchers are plainly attired in shades of olive green, gray, or brown, often with yellowish or whitish underparts. A bolder pattern, exhibited with minor variations by no less than eight species in five genera, might be called the kiskadee type, from its resemblance to this large, far-ranging bird. In these species the black or dusky head bears prominent white superciliary stripes that converge on the back of the head. The upper parts are brownish or olive, the underparts bright yellow, usually with a white throat. Largely or almost wholly black or blue-black are Crested Black-Flycatchers, Velvety Black-Flycatchers, Blue-billed Black-Flycatchers, Spectacled Flycatchers, and Black Phoebes, all but the last confined to South America. The male White-headed Marsh-Flycatcher is black with a white head. The Pied Water-Flycatcher is black and white, with the white more extensive, covering most of the head and the underparts. The White Monjita, another South American species, is black only on the primary feathers and tip of the tail.

A few flycatchers are more warmly colored. The male Vermilion Flycatcher glows with scarlet crown and underparts, contrasting with his dark upper parts. An exceptionally ornate member of the family is the Many-colored Rush-Flycatcher—white, yellow, red, green, blue, and black in an

White Monjita, *Xolmis irupero*. Sexes alike. Eastern Brazil to central Argentina.

elaborate pattern. The sexes of this elegant little bird of South American marshes are alike, as is true of nearly all flycatchers. Much more frequent than bright spectral colors are attractive shades of cinnamon, ochre, rufous, or chocolate, as on the Cliff Flycatcher, Streaked-throated Bush-Flycatcher, Rufous-breasted Chat-Flycatcher, and Cinnamon Flycatcher and less extensively on many others, especially on wings and tail. The softly blended shades of bush-flycatchers *(Myiotheretes)* are lovely.

On the center of the crown of many flycatchers are spots of white, yellow, or some shade of red that are hidden except when the crown feathers are spread in moments of excitement, as in territorial encounters or courtship. A few species have conspicuous crests. The Crested Black-Flycatcher of Brazil wears a bushy tuft of pointed feathers. The crest of the Tufted Tit-Flycatcher of the Andes is composed of long, narrow, recurved black feathers. The swept back crest of the Tufted Flycatcher of Central American mountains resembles that of the North American Tufted Titmouse. With its crown feathers laid flat, the widespread Yellow-bellied Elaenia is often difficult to distinguish from several similar species. When, in a questioning or antagonistic mood, the little gray bird erects these feathers in a double-peaked crest

Tufted Tit-Flycatcher,
Anairetes parulus. Sexes
alike. Colombia through
Andes to Tierra del Fuego.

with a patch of white in its midst, this elaenia is unmistakable.

The most splendid of crests adorns the head of the Royal Flycatcher, which frequents forest streams from Mexico to Bolivia. When laid flat, the crest projects behind the crown as a little tuft of feathers, imparting an odd, hammerheaded appearance, and giving no hint of its elegance. All too seldom, it is raised and spread fanwise, encircling the head with a wide scarlet aureole, each feather of which is tipped with a spot of velvety black and bluish violet. The female's crest is similar but paler. In many hours of watching Royal Flycatchers at their nests, I have only rarely seen them spread their crests, and never on casual meetings in the forest. But one time in Honduras I enjoyed an unforgettable display of their diadems while I took shelter from a tropical downpour beneath a ledge of rock, overgrown with mosses and ferns, beside a mountain torrent. While I vainly tried to keep dry, the pair of flycatchers I was studying welcomed the shower, flitting from twig to twig above the stream, spreading wings and tail to the raindrops, and preening their plumage with evident satisfaction. In this general rearrangement of feathers, their crests were not neglected. First the male, then his consort, spread their gorgeous tiaras in a rev-

Long-tailed Flycatcher, *Colonia colonus*. Male, female similar but shorter tailed. Northeastern Honduras to western Ecuador, the Guianas, and northeastern Argentina.

elation of splendor that I had not suspected. All too soon, the crests were folded and the birds resumed their undistinguished, hammer-headed aspect.

Wattles and areas of bare skin on head or neck, frequent in the related cotingas, are rare in flycatchers. The Spectacled Flycatcher of southern South America is named for the ruffled yellow wattle around each eye. The Strange-tailed Flycatcher, to which we shall presently return, displays in the breeding season bright orange bare sides of the head, the chin, and the throat.

A few flycatchers wear long pennants in their tails. The

Strange-tailed Flycatcher,
Alectrurus risoria. Male.
Southern Brazil and
Paraguay to central
Argentina.

bifurcated tail of the Streamer-tailed Flycatcher is composed
of graduated, tapering feathers, the longest of which exceed
the body in length. Almost equally long are the ribbonlike
outer feathers of the forked tails of Scissor-tailed and Fork-
tailed flycatchers. The central tail feathers of the small Long-
tailed Flycatcher expand from a long, narrow base to
broader ends. At the base of the Strange-tailed Flycatcher's
long outer tail feathers the shafts are bare; at their center the
inner web broadens, then tapers to a filamentous tip. The
outer web is lacking. Most curious of all is the tail of the
Cock-tailed Flycatcher, well described by W. H. Hudson:
"The two outer tail-feathers have remarkably stout shafts,
with broad, coarse webs, and look like stumps of two large
feathers originally intended for a bigger bird, and finally cut
off near the base and given to a very small one. In the male

Cock-tailed Flycatcher,
Alectrurus tricolor. Male.
Bolivia and southern
Brazil to northeastern
Argentina.

these two feathers are carried vertically and at right angles
to the plane of the body, giving the bird a resemblance to a
diminutive cock; hence the vernacular name *Gallito*, or Little Cock, by which it is known" (1920, I:150).

The bills of flycatchers are diverse, corresponding to
their different ways of foraging. In length they range from
the heavy beaks of the Boat-billed Flycatcher and the Great
Shrike-Flycatcher, nearly an inch and a quarter long, to tiny
bills barely three-eights of an inch in length, as in the
Ruddy-tailed and Long-tailed flycatchers. The bills of many
flycatchers taper from a broad base to a point, with a diminutive hook at the tip of the upper mandible. The short bills
of the spade-billed flycatchers and the flatbills are about as
broad at the base as they are long; viewed from above, they
are almost equilateral triangles. Contrasting with these, the
tody-flycatchers' bills are long and flat, with nearly parallel
sides. Some of the smaller flycatchers, such as the Ochre-
bellied and the Mistletoe, have slender, warblerlike bills. Although most flycatchers' bills are straight, that of the bentbills is downcurved.

From the skin at the base of the bill of most flycatchers
spring bristles that are modified, hairlike feathers, often
with barbs at the base. Those around the edges of the mouth,

where they are especially prominent, are known as rictal bristles. In length they vary greatly; on the Sulphur-rumped Flycatcher they are about as long as the bill; on the Royal Flycatcher they are only slightly shorter; from these they range downward in length to those of the beardless fly-catchers (*Camptostoma*), so-called because their bristles are rudimentary. How rictal bristles serve the birds that wear them is far from clear. One suggestion is that they help fly-catchers capture flying insects by deflecting to their gapes some that might be missed by the bill. Weighing against this view is the fact that rictal bristles are worn by birds that rarely capture flying insects, such as American Robins and Brown Thrashers, but absent or too short to be of much help on some aerial flycatching birds, such as swallows: there is little correlation between the development of rictal bristles and mode of foraging. Moreover, high-speed photography has revealed that Great Crested Flycatchers, Eastern Phoebes, Eastern Wood-Pewees, and at least some species of *Empidonax* seize flies in the tip of the bill, not in the mouth. And Willow Flycatchers whose bristles had been cut off short showed no loss of ability to catch flies. The suggestion that the facial bristles of birds have a tactile function, like the vibrissae of mammals, has not been proved. Experiments in a wind tunnel supported the idea that rictal bristles protect the eyes from insects that the bird has failed to capture, or from fragments detached from prey; much as the narial bristles springing from the forehead shield the nostrils from small particles.

Many of the larger flycatchers, and not a few of the smaller ones, are readily identified; but the host of diminutive species without bright colors or bold patterns puzzle not only amateur birdwatchers but likewise professional ornithologists. To identify them one must give close attention to subtle differences in coloration and to such features as shape of bill, eye rings, wing bars, and feather edgings. After one becomes familiar with voices, they are the greatest aid to identification; sometimes two species that look

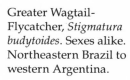

Greater Wagtail-
Flycatcher, *Stigmatura
budytoides*. Sexes alike.
Northeastern Brazil to
western Argentina.

too much alike to be distinguished by appearance can be separated by their notes. Many small flycatchers of undistinguished aspect build unusual nests that help to reveal their identity. Often a nest unlike any other I had ever seen has brought awareness of some small, unobtrusive flycatcher of forest or thicket that I might otherwise have overlooked. Many a tiny flycatcher compensates for its plainness by building a wonderful nest.

Birds so dependent upon active insects as flycatchers migrate when winter's chill kills or immobilizes their prey. All but a few of the great host of flycatchers that spread over forests and fields of the United States and Canada vanish as the cold season approaches, many departing as early as August. Most appear to travel at night, in company with throngs of wood warblers, vireos, thrushes, and other small birds. Not only do birds migrating by night appear to be less exposed to predation by raptors than they would be by day, but after daybreak they can rest and forage to replenish energy spent on their long nocturnal flights. Small birds that migrate in the daytime forage along the way, in the manner of swallows and a few flycatchers. Among the diurnal migrants are Eastern Kingbirds. In the early morning, in late afternoon, and in cloudy weather, they often pass overhead at no great height, southward in autumn, northward in spring, in small parties or flocks of a hundred or more. At intervals they alight in exposed treetops, from which they dart out to capture insects or to pluck berries from trees.

While passing through the long Central American isthmus between the United States and South America, flycatchers may interrupt their journeys at convenient points. Through most of September and early October of 1930, many Eastern Kingbirds roosted nightly in a large patch of tall elephant grass in a cleared valley on the north coast of Honduras. Here, the most silent members of a chattering multitude, they slept with scores of resident seedeaters, grassquits, and a few orioles. Similarly, for nearly a month, from mid-April to the first quarter of May 1936, northward-bound Eastern Kingbirds roosted every night in low trees on a tiny island in the Río Buena Vista, a mountain torrent in the foothills of the Cordillera de Talamanca in southern Costa Rica. Whether the same individual birds interrupted their journeys for the long intervals that these roosts were occupied, I could not tell; the roosts may have been the resorts of a succession of migrants, but evidently at least a few individuals passed several nights in each of them and led new arrivals to them.

Western Kingbirds also migrate by day. In January and early February of 1963, Burt L. Monroe, Jr., watched moderate numbers of these flycatchers pass daily over the Gulf of Fonseca, which lies between Nicaragua and El Salvador, all flying in the same northerly direction. In the same region, on February 1, about four thousand Scissor-tailed Flycatchers flew northward over the gulf from dawn to dusk. They also migrate by night; their unmistakable silhouettes have been seen passing over the face of the moon.

Like other North American migrants, most flycatchers go no farther than Mexico, Central America, and less often the West Indies; but some undertake longer journeys, passing on to South America, perhaps leaving a few stragglers of their species in these intermediate regions. One of the greatest travelers is the Eastern Kingbird, which from Canada and the United States reaches northern Chile and Argentina, although many individuals go no farther than Colombia and Venezuela. Other northern flycatchers that winter chiefly in South America are the Olive-sided, Eastern Wood-Pewee, and Western Wood-Pewee, some of which travel as far as southern Peru, Bolivia, and Brazil. Among the hardiest of the northern flycatchers are the phoebes, which do not migrate as far southward as many others. An occasional Eastern Phoebe braves snow and ice as far north as New England and southern Ontario. Many Say's Phoebes, some of which nest north of the Arctic Circle in Alaska, pass the winter in the western United States or go no farther than central Mexico. Black Phoebes winter in California, where they breed. Versatile foraging, including gleaning from the ground, helps them to avoid starving on cold days when few insects fly.

Bird migration is more prominent in North America, where most of the land lies in the North Temperate Zone and the Arctic, than in South America, which spreads broadly across the tropics and narrows southward without reaching the Antarctic. Many of the flycatchers that breed in the more southerly parts of Chile, Argentina, and even Brazil move northward in the austral winter, while those in the southern Andes descend to lower altitudes if they do not migrate to the north. At least six species of ground-flycatchers (*Muscisaxicola*) join this northward movement. Populations of the White-crested Elaenia that nest in southern Argentina and Chile, down to Tierra del Fuego, fly northward to join resident populations, apparently as far as the highlands of Colombia. Many flycatchers that breed in Patagonia move from the southern tip of the continent to central Argentina; Rufous-backed Negritos to Uruguay, Paraguay, Bolivia, and southeastern Brazil. Some of the birds that migrate from the higher latitudes of South America mingle with members of their species that reside permanently at lower latitudes, making it difficult to trace the migrants' movements. Some-

Rufous-backed Negrito,
Lessonia rufa. Male. Peru and
Bolivia to Tierra del Fuego,
migrates northward in
austral winter.

times migrants can be distinguished from the residents they join by their wings, more pointed for long-distance flight.

A few species of birds that breed widely in South America contain populations that visit the northern continent to nest, then return to their southern homeland. Conspicuous among them is the Piratic Flycatcher, a vociferous bird resident in much of South America, from which some spread over Central America and southeastern Mexico in February and March, breed in structures stolen from other birds, then in the autumn return to the southern continent. A somewhat similar pattern is followed by the Streaked Flycatcher but without detriment to the resident birds of Middle America. No passerine bird is known to migrate southward from North America to breed.

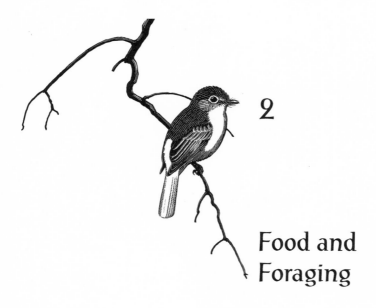

2

Food and Foraging

Perching upon a high exposed twig or overhead wire, turning its head from side to side, a flycatcher watches sharply for passing insects. Suddenly it darts up or out or down, pursues its flying prey, seizes it with a snap audible if one is near enough, and flies back to the same or another perch, against which the predator may beat the victim before swallowing it. The dashing pursuit may be short or long; often the bird rises high above the treetops, to twist and turn following the movements of an insect. In the evening, when after a warm day the cooling air is full of insect life, I have watched Tropical Kingbirds soar high, to circle and hover, rise and fall, snatching up insect after insect without a pause, swallowlike, before they dropped gracefully back to a perch.

With longer or shorter range, aerial hawking is the principal mode of flycatching for kingbirds, wood-pewees,

Streamer-tailed
Flycatcher, *Gubernetes
yetapa*. Sexes similar.
Bolivia and southern
Brazil to central
Argentina.

Olive-sided Flycatchers, Long-tailed Flycatchers, and Cliff
Flycatchers, and it is practiced more or less frequently by
many others, large and small. Because it is so spectacular
that it claims attention, we might conclude that this method
of foraging prevails throughout the family—and we would
be wrong. Only a minority of the species forage mainly in
this fashion, and how frequently they employ it depends
upon the weather and other circumstances; as temperature
drops and cloudiness increases, Eastern Kingbirds make
fewer aerial sallies and more often gather food from foliage
or the ground. Many flycatchers procure their food in ways
that are less eye-catching and surprisingly diverse, including much fruit eating.

Instead of hawking for insects under the open sky, some
woodland flycatchers pursue them less conspicuously. A
treefall in a montane forest makes space enough for Tufted-

Flycatchers' aerial foraging. Perching well above the ground but far below the crests of the taller trees, pairs of these sprightly little birds make frequent sallies to capture tiny flying insects, returning again and again, day after day, to their exposed lookout in the same small forest opening. As a Sulphur-rumped Flycatcher wanders through the rain forest with a flock of diverse species or joins a party of birds foraging with army ants, it weaves airily through the lower boughs of trees and higher shrubs, snatching insects from the air or plucking them from leaves. Amid low, bushy growth, little Bran-colored Flycatchers capture insects while they flit across narrow open spaces.

Instead of pursuing their prey in the air, many small flycatchers and some larger ones gather spiders, caterpillars, and stationary winged insects from foliage and branches. Most frequently these birds snatch prey from the undersides of leaves, where in tropical rain forest insects tend to be more abundant than on the upper faces. Moreover, the birds appear to enjoy an ampler view looking upward from the leafy twig on which they rest, which may screen much of the foliage beneath them. By a short upward dart, perhaps followed by a brief hover, the bird secures its prey before dropping to another perch. Among the many flycatchers that obtain a major part of their food by this method are the Acadian, Royal, Yellow-olive, and Ruddy-tailed, the Black-capped Pygmy-Flycatcher, Golden-crowned Spadebill, and tody-flycatchers. The bentbills' slightly downcurved beak appears to be specially adapted to this mode of foraging. As the tiny bird darts obliquely upward against the underside of a leaf, the bill's tip would be more nearly parallel to the leaf's surface than if it were straight, and should slip more readily beneath a caterpillar or other small creature clinging there.

An unusual method of foraging is practiced by White-ringed Flycatchers. In the Caribbean lowlands of Costa Rica, I found these middle-sized, yellow-breasted flycatchers traveling over the rain forest in groups of two to four. Perching conspicuously on the ends of high, exposed leafy twigs, often the topmost shoots of towering trees, they scanned the foliage around and below them. When they detected a caterpillar, mature insect, or spider on a branch or the upper surface of a leaf, they flew out, or more frequently down, to pluck it off. Often they shifted from one high perch to another to bring fresh foliage into view. Before swallowing a large victim, they knocked it against a branch. These White-ringed Flycatchers rarely caught flying insects.

Some small flycatchers forage like warblers and vireos. With short, uptilted tails wagging up and down, beardless flycatchers, both Northern and Southern, hop and flit through the branches of trees and shrubs, often high but sometimes low, gleaning minute creatures from foliage, twigs, or flowers. Occasionally they gather something from a solid surface while hovering on wing, but they rarely catch flying insects, an omission that may be related to their lack of more than rudimentary rictal bristles. Their diet includes ants, spiders, and small berries. Similar perch-gleaning is practiced by White-crested Elaenias, Suriri Flycatchers, and Common Tody-Flycatchers, which sidle along twigs wagging their narrow tails rapidly from side to side.

A method of foraging frequent among flycatchers is ground-gazing. The bird perches low, on a stump, fence wire, rock, stout weed stalk, or low branch, closely watching the ground. When it spies an insect or some other edible creature, it drops down, seizes the prey, and returns to its low stand. Black Phoebes often forage in this manner, gathering small items from a lawn or from bare ground beside a stream. In wide expanses of grass with scattered low bushes, I have found Fork-tailed Flycatchers perching for hours on them, to drop down to the grass or ground for prey—which seemed a strange way for birds with such long streamers in their tails to forage. In our Costa Rican garden, wintering Yellow-bellied Flycatchers often use this method. W. H. Hudson told how, on the Argentine pampas, ground-gazing Pepoaza Flycatchers dropped from the summit of a stalk or bush, or even a low tree, to seize terrestrial beetles or grasshoppers in their feet—a most unflycatcher-like method. Then, swaying about and opening its wings to keep its balance, the flycatcher tears an insect apart between foot and bill and devours it before returning to its lookout. Another ground-gazer of South America is the White-rumped Monjita; and in the Andes, shrike-flycatchers often forage in this manner.

More terrestrial are the ground-flycatchers of the genus *Muscisaxicola*, so named for their superficial resemblance to Old World chats. High in the Andes and at lower elevations in southern South America, they walk and run over bare rocky slopes or more level ground, often not far from a stream or pond, picking up insects and other small invertebrates. Equally terrestrial is the Rufous-backed Negrito, which from Peru to Tierra del Fuego prefers damp, open ground, at moderate elevations or by the southern coast. Over much of South America, Masked Water-Flycatchers gather nourishment from the surface of marshes and river banks. From Venezuela to central Argentina, short-winged, long-legged Cattle Flycatchers find nearly all their food on open ground, over which they run with a speed unexpected in a passerine bird—so swiftly that, as Hudson affirmed, they can overtake and capture flying insects without rising into the air. Between terrestrial pursuits, they commonly rest on and watch from the backs of horses, donkeys, cows, or other quadrupeds, earning for themselves the name *jinete* (horseman) in Venezuela. Instead of depending, in the manner of anis and Cattle Egrets, upon grazing mammals to stir up insects from the grass, Scissor-tailed Flycatchers in Texas accompany flocks of Río Grande Turkeys, seizing in the air grasshoppers and other insects that the turkeys flush while foraging.

Skillful foragers in perilous situations are the small, pale gray and blackish Torrent Flycatchers, which frequent swift, boulder-strewn mountain torrents or sometimes calmer streams from Costa Rica to Bolivia. They forage

Examples of ecological convergence.
Above left, European Whinchat, Saxicola rubetra (thrush family). *Above right*, Plain-capped Ground-Flycatcher, *Muscisaxicola alpina*.
Below left, Sulphur-rumped Flycatcher, *Myiobius sulphureipygius*. *Below right*, American Redstart, *Setophaga ruticilla* (wood warbler family).

chiefly from rocks that rise in the channel or line the shore, rarely penetrating streamside vegetation. It is absorbing to watch these birds flit airily from boulder to boulder, frequently rising above the waterway to twist and turn with graceful skill while they catch flying insects, then darting softly to seize a caterpillar or spider from foliage that overhangs the stream. They stoop to pick a morsel from the boulder where they stand, or hover momentarily before the vertical or overhanging face of some taller rock to snatch up a

Torrent Flycatcher,
Serpophaga cinerea.
Sexes alike. Costa Rica
to northern Bolivia.

minute creature crawling there. They hop lightly over the flat surfaces of wider ledges and skim easily across smoother reaches of the stream, almost touching the water. They cling to the steep, wet sides of rocks overgrown by seed plants (Podostemonaceae) that resemble tiny mosses or small green algae, among which they find much to eat.

　　Often the Torrent Flycatchers alight upon a rock over which a thin sheet of water flows, picking inconspicuous items from its glistening face. As they stand with their feet in the water, their breasts are wetted by the surge. With breathtaking daring, they pluck inviting morsels from the very brink of white-water rapids, where their plumage is wetted by the spray and they seem in danger of being sucked under by lapping waves, which they skillfully avoid, flitting away just in time to save themselves. Ever active, seeming as light in their movements as downy feathers wafted by the breeze, they hunt silently and swiftly, catching chiefly creatures too small to be detected by a watcher on the shore. Often they capture small gray moths, and occasionally they take a dragonfly as long as themselves, which they gulp down with difficulty. While standing on the rocks looking for insects, they constantly wag their short black tails up and down, in the manner of many fluviatile birds.

The diet of many flycatchers consists wholly, or almost wholly, of larval and adult insects, spiders, millipedes, scorpions, and other small invertebrates. Examination of the stomach contents of such aerial flycatchers as the Eastern Wood-Pewee, Western Wood-Pewee, and Olive-sided Flycatcher revealed from a trace to about 1 percent of vegetable matter. The Black Phoebe, with a quite different method of foraging, takes no more. Among the small tropical flycatchers that I have watched, I have never seen a Sulphur-rumped, Long-tailed, or Torrent flycatcher eat a berry. Stomachs of Yellow-bellied and Least flycatchers contained 2 or 3 percent vegetable matter. Berries of a number of trees and shrubs accounted for 6.3 percent of Great Crested Flycatchers' diets, and 10.8 percent for Eastern Phoebes. In temperate North America, flycatchers tend to eat more fruit in autumn, winter, and early spring than in summer when they nest.

As recompense for scattering their seeds, tropical trees, shrubs, and vines offer birds a great variety of berries, one-seeded drupes, and arillate seeds. The last-mentioned, little known in northern lands, are wholly or partly covered by soft edible tissue—the aril—overlying a seed coat that is usually hard or at least indigestible. Such seeds are borne, singly or many together, in a tough or woody inedible pod, often red or yellow. The pod does not open until the seeds are mature, ready to be swallowed by the birds. After digesting the arils, birds regurgitate or defecate the still-viable seeds. Usually rich in oil but poor in starch and sugar, brightly colored arils are eagerly sought by birds as soon as the opening pod exposes them to view. Flycatchers are prominent among the host of tropical birds that profit from the rich sources of nourishment plants provide.

Berries and arillate seeds are important in the diet of Ochre-bellied Flycatchers, small, exceedingly plain inhabitants of tropical woodlands. In Trinidad, David and Barbara Snow saw them take the fruits and arillate seeds of ten species of trees and shrubs, and in Costa Rica I have recorded the same number in their fare. Another bird with a largely vegetarian diet is the tiny gray Mistletoe Flycatcher, disparagingly called the Paltry Tyrannulet. In addition to taking the drupes of several mistletoe species that infest tropical trees and feeding these fruits to their nestlings, Mistletoe Flycatchers eat berries of other kinds, often plucking them by darting up and hovering beside a cluster. Other species that eat many berries and arillate seeds are the Vermilion-crowned, Gray-capped, Piratic, and Boat-billed flycatchers and the Yellow-bellied and Lesser elaenias. Even an aerial hawker like the Tropical Kingbird supplements its diet with many berries and arillate seeds.

Whether a flycatcher gathers its fruits while perching beside a cluster, hovering beside it, or darting past appears to depend more upon the position of the fruit and the strength of its attachment than upon the bird's anatomy. Small fruits are nearly always swallowed whole. Sometimes a flycatcher tries to gulp down a fruit too big for it. Rich in digestible oil, the hard green drupes of the Aceituno tree, resembling miniature olives, are eagerly sought by birds as diverse as chachalacas, motmots, and thrushes. A Gray-capped Flycatcher that tried to swallow them found them too big. After each failure to force a fruit down, the bird knocked it against the perch as though it were an insect; but the aceituno drupe was too hard to be softened by this treatment. Several times the pressure of the bird's mandibles made the fruit shoot out of its bill, as one can project a small slippery object by pressing it between a thumb and finger. When this occurred, the flycatcher darted forward and caught the drupe before it had gone the bird's own length—an amazingly prompt reaction.

Some flycatchers, especially small species such as the Yellow-olive Flycatcher and Northern Bentbill, which subsist chiefly upon insects caught amid foliage, are foraging specialists, eating a limited range of items procured in few ways. Likewise the Cliff Flycatcher appears to depend almost wholly upon aerial hawking for its food. Other flycatchers take a wide range of foods by very diverse means. Outstanding among these generalists are Vermilion-crowned Flycatchers, common birds from northeastern Mexico to northern Argentina. To catch flying insects, they sally far upward or outward, then return to the tree from which they started. Often they ground-gaze on a rock, fence wire, or low branch in a pasture, from which they drop to the ground to seize an insect or spider. Or they hop over short-cropped grass, gathering small items. Those dwelling near rivers or ponds hunt over moist bare ground at the water's edge or upon bare rock ledges. They venture into shallow water up to their thighs to catch small tadpoles. From a perch above deeper water, they drop to pick some floating edible object deftly from the surface, without submerging themselves. They take many kinds of fruits, gathering the small red drupes of royal palms while hovering beside a heavy cluster, swallowing whole the lead-colored drupes of the Aguacatillo tree, plucking berries of numerous species of trees and shrubs of the melastome family, eating mistletoe drupes, and taking at least five kinds of arillate seeds from trees and vines around our house. A complete list of the Vermilion-crowned Flycatcher's foods, vegetable and animal, would be long.

Another versatile forager is the Bright-rumped Attila, a bird almost as widespread in tropical America as is the Vermilion-crowned Flycatcher but less often seen. From the day they hatch, attilas are fed with lizards, at first tiny but increasing in size as the nestlings grow. Small frogs and a few insects fill out their fare. Adults often forage on open ground, dropping from stumps or logs to seize their prey, or even hopping with feet together over close-cropped grass while they pick up insects. Sometimes they join a motley crowd of birds that follow army ants, overtaking with dashing flight any insects that try to escape the ant horde. Arillate seeds vary their diet.

The most nearly omnivorous flycatcher that I know is the Great Kiskadee. The equally big Boat-bill is satisfied with insects, often big cicadas, and small fruits, but the kiskadee's appetite is boundless. It eats many berries and drupes of palms, and in the great banana plantations of warm lands is so eager for this fruit that I have heard the kiskadee called "the banana bird." Hot peppers impart piquancy to its fare. Besides fruits, cooked rice or bread and milk attract these birds to feeding trays. To satisfy their craving for animal foods, they adopt the procedures of a diversity of birds. Like a kingbird, the kiskadee captures moths, beetles, wasps, and many other insects on the wing. From an overhanging bough or projecting rock, it plunges into the water to catch small fishes, like a kingfisher if less expertly. Like a heron, the kiskadee wades in shallow pools for tadpoles, wetting its belly and whole head, which it must immerse to seize victims with its short bill. Like a ground-gazer, the kiskadee rests on a branch or wire, its big head cocked sideways to peer intently down until it spies a grasshopper, lizard, frog, or small snake, perhaps even a mouse. Then it dashes down upon the prey. Where such perches are lacking, as on open grassy plains, the kiskadee ranges like a kestrel, searching for prey below. Having seized an insect or larger victim, the bird beats this resoundingly against a perch to extinguish its life and make it swallowable. Mice require prolonged treatment before a kiskadee can force them down. Snail shells are broken in the same manner.

Like a Black Vulture, the Great Kiskadee scavenges for stranded fishes and other offal on the seashore, and it also frequents garbage dumps. Hudson recorded that in spring on the pampas, these flycatchers followed the plough in company with gulls, cowbirds, Guira Cuckoos, and other birds, hopping awkwardly over the rough, upturned earth to pick up worms and larvae, since kiskadees cannot run like more terrestrial flycatchers. The same writer also saw kiskadees following rural butchers' carts, waiting for an opportunity to dash in and carry off any scrap of meat or fat they were able to detach. As though all these sources of

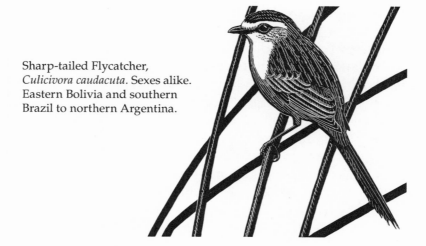

Sharp-tailed Flycatcher,
Culicivora caudacuta. Sexes alike.
Eastern Bolivia and southern
Brazil to northern Argentina.

nourishment were not sufficient, the kiskadee devours nestlings and fledglings of smaller birds, as I have myself witnessed. Probably no individual kiskadee has foraged in all these ways, but some use many of them, as opportunity arises. In their ways of foraging, flycatchers range from rather narrow specialization to extreme flexibility, with the Great Kiskadee the most versatile of all.

Great Kiskadees are not the only fish-catching flycatchers. Both the Black and the Eastern phoebes occasionally engage in this activity. In Chicago, Laurence Binford watched two of the latter drop repeatedly from a low concrete wall and pluck small fishes from just below the surface of still water, without wetting more than their foreheads, bills, throats, and chests. On the West Indian island of Barbados, two Gray Kingbirds swooped down from perches to seize small fishes struggling to cross an exposed sandbar that blocked their ascent of a small stream; the birds then carried their victims back to their perches to pound and devour them.

Differences in structure adapt flycatchers to their diverse modes of foraging. Aerial hawkers have triangular bills, very long wings like long-distance migrants, and relatively short legs. Flycatchers that dart upward to pluck insects from beneath leaves have broad bills, short wings, relatively long legs, and reduced tails. Both perch-gleaners and ground-foragers have narrow bills, and the legs of the latter are long. In generalists or versatile foragers, these features are intermediate. Apparently, specialists finely adapted to one mode of foraging have evolved from more generalized ancestors.

3

Daily Life

Most flycatchers breed in monogamous pairs. Migration disrupts these pairs, but permanent residence favors their maintenance throughout the year. Among the dozen tropical flycatchers that I have found in pairs at all seasons are the Tropical Kingbird, Gray-capped and Vermilion-crowned flycatchers, Yellow-bellied Elaenia, Common Tody-Flycatcher, Torrent Flycatcher, and Golden-crowned Spadebill. Tropical birds of many families live in pairs throughout the year, and many of these birds probably retain the same partners as long as both survive. Well beyond the tropics, in central Argentina where some flycatchers are only summer residents, Hudson found permanent residents in pairs through the winter months, among them the Great Kiskadee, Cattle Flycatcher, and Sooty Flycatcher, his "Little Riverside Grey Tyrant." Among tropical flycatchers are a small number of species that never form pairs, the females attending their nests unaided. To these we shall return in chapter 6.

In their winter homes in tropical America, migratory *Empidonax* flycatchers lead solitary lives. In early September of some years, sharp little notes tell me that two Yellow-bellied Flycatchers have arrived from their far-northern nesting grounds and are contending for sole occupancy of our garden. Soon one vanishes, leaving the other in undisputed possession of the territory, where until late in the following May we shall see it, silently perching low in the shrubbery, sallying forth to pluck insects from foliage or the ground, rarely eating a berry. Least Flycatchers also defend individual territories during their long sojourn in the tropics, as do many wintering wood warblers and birds of other families.

Despite their separation during the winter, a pair of Willow Flycatchers banded in Michigan by Lawrence Walkinshaw nested together for four years of the five that both survived. Their philopatry, or strong attachment to the territory where thay had successfully nested together, was probably responsible for this fidelity. Even Black Phoebes permanently resident in central California separate from their mates after the breeding season and each holds an individual territory through the colder months, thereby compensating for reduced abundance of food by spreading out more widely. In the tropical parts of the Black Phoebe's extensive range, however, I have frequently found two together during the months when they do not breed.

After nesting in southern Patagonia, Chocolate Flycatchers pass the austral winter on the Argentine pampas, where Hudson found them foraging over the ground in flocks of half a dozen to twenty individuals; but except while migrating by day, flycatchers are rarely found in flocks of a single species. After the breeding season, residents forage alone, in pairs, or with their self-supporting young. A few regularly accompany the mixed-species flocks of tropical woodlands, where one can wander long, disappointed in the expectation of seeing many birds, until one meets one of these motley crowds and finds among the trees or under-growth more flitting figures than can be kept in view. Low-land flocks, composed largely of antbirds, ovenbirds, wood-creepers, and tanagers, contain only a minority of flycatchers.

In the very large mixed flocks in the bird-rich Amazon-ian forests of eastern Peru, pairs of Yellow-margined Fly-catchers are among the six nuclear species about which many others gather; the other five belong to different fami-lies. Frequent attendants of these flocks include Forest Elae-nias, Yellow-crowned Flycatchers, and White-lored Fly-catchers. In the humid lowland forests of Central America, a mixed-species flock often includes a single Sulphur-rumped Flycatcher, fanning its tail and drooping its wings in a sort of perpetual display as it snatches insects from fo-liage or the air.

Amid oak trees and pines in the highlands of southern Mexico and northern Central America, flocks composed largely of wintering and resident wood warblers often in-clude a pair of Tufted Flycatchers and a single Greater Pewee. Should a second pair of Tufted Flycatchers or a sec-ond pewee try to join the flock, a noisy confrontation arises and continues until either the intruders or the original mem-bers are expelled. In addition to their constant attendants, mixed flocks everywhere are often joined by other flycatch-ers that accompany them for a while, then drop out. Among the advantages of foraging in a flock is the increased secu-rity from predators gained by moving amid many vigilant companions. Although the flock is certainly more conspic-uous than a solitary individual, the number of watchful birds compensates for the attention that it attracts.

Among the crowds of birds of diverse kinds that gather over and around a swarm of army ants, snatching up insects and other small creatures that the hunting horde drives from concealment in the ground litter or from crevices in the bark of trees, flycatchers are occasionally present. None that I know is a regular follower of army ants; but a Golden-crowned Spadebill, a Bright-rumped Attila, or a Sulphur-rumped Flycatcher, meeting one of these swarms as it flows over the forest floor, may remain on its outskirts long enough to catch a few escaping insects, then go its own way. More than many resident species, wintering Acadian Fly-catchers take advantage of the activities of army ants. They prefer the smaller swarms of *Labidus praedator* to the more numerous and active *Eciton burchelli,* which draws larger numbers of resident birds. Migrants that come from afar to the tropics have a marvelous ability to adjust promptly to conditions very different from those in the northern lands where they grew up.

Flycatchers bathe both in standing water and above ground, probably more often the latter. M. Slessers, who for long hours watched appropriate ponds from camouflaged blinds, saw Eastern Wood-Pewees and Eastern Phoebes bathe once or twice on a summer day. Two phoebes, who remained in Maryland at least until January, when snow covered the ground and ice formed over the pools, bathed once or twice a week in small patches of open water. Instead of partly immersing themselves in shallow water and splashing it over themselves with their wings, as many small birds do, flycatchers dive to water from a perch, skim over it just deeply enough to raise a spray and scoop the liquid over their backs, then continue their flight to a perch where they shake their wet feathers. They may repeat this maneuver, reminding one of bathing swallows or swifts. This method of bathing becomes spectacularly beautiful when practiced by Scissor-tailed Flycatchers in Oklahoma. Margaret Nice told how, from willows growing by a small pond, a dozen or more of these long-tailed birds swept one by one over the water without quite touching it. Finally, a bolder bird splashed its breast into the water, to be followed by all the others, each dipping momentarily into the pond before rising again.

Other flycatchers bathe by sliding over a large leaf after rain, breast pressed against it to gather the glittering drops in their plumage, then vigorously vibrate their wings, as I have watched a Lesser Elaenia do. Still others simply permit

White-headed Marsh-Flycatcher,
Arundinicola leucocephala.
Male. South America from
Colombia to northern Argentina.

falling rain drops to enter their plumage while they preen,
as a pair of Royal Flycatchers did on the memorable occasion
when I first saw them spread their magnificent crests
(chapter 1).

Flycatchers appear seldom, if ever, to dust themselves
or "dust bathe." Once, while I sat in a blind in front of a trogons'
nest in a Panamanian forest, a Dusky-capped Flycatcher
sunned itself in a patch of sunshine that fell upon a
prostrate trunk close beside me. Lying flat, the bird spread
its wings and fluffed up its body feathers to permit the rays
to penetrate deeply into them. A short while before, a hummingbird—a
Crowned Woodnymph—had sunned itself in
the same spot in much the same attitude.

When flycatchers chase other birds, of the same or a different
species, without any evident motive and without
harming them, they appear to be playing. As night fell over
a tree where many Scissor-tailed Flycatchers roosted, Frank
W. Fitch, Jr., watched them indulging in aerial games of tag,
chasing one another over and around the tree. These pursuits
often ended in brief encounters in which neither participant
was hurt; they were less intense than clashes that
occur during pair formation and territorial defense. One
morning in April, I saw a Boat-billed Flycatcher with a big

brown feather in its bill alight in a tree. The Boat-bill struck
the feather against a perch, as though it were an insect. After
a while the bird dropped the plume; and as it floated slowly
downward, the mate of the first bird darted out and caught
it, then alighted and knocked it against a branch. Soon the
second bird released the feather, only to shoot out and seize
it as it wafted away, then beat it against the branch again.
Carrying the feather to a neighboring tree, the bird continued to beat it but more gently than it would have done to a
large insect. Finally, the Boat-bill with the feather let it fall,
and the pair flew away. They were playing with the feather.

As day ends, birds seek their roosts. Many of the smaller
flycatchers of woodland and thicket hide themselves so well
amid the verdure that they are seldom discovered while
they sleep. Some of the larger ones of more open places are
more readily found. Tropical Kingbirds, Gray-capped Flycatchers, Vermilion-crowned Flycatchers, Boat-billed Flycatchers, and Yellow-bellied Elaenias often sleep in a thorny
orange tree, or in a shade tree in a pasture, where their yellow breasts are readily picked out by a flashlight's beam.
Frequently they roost upon a lower branch, shielded from
the view of owls and large predatory bats by the leafy
crown above them, but with a clear space below into which
they can readily drop to fly away if a prowling mammal or
snake, approaching along the branch, wakes them by shak-
ing it. Occasionally they roost where they are exposed
above as well as below. Mates sleep from a few feet to a few
yards apart, as tanagers also do.

Fledgling Tropical Kingbirds, Vermilion-crowned Flycatchers, and Common Tody-Flycatchers sleep in closest
contact in a row, an arrangement that may be continued
for weeks or months. Sometimes a parent tody-flycatcher
rests beside its fledged young in a row of three, whose
breasts make one continuous expanse of yellow. In Panama,
three Rusty-margined Flycatchers went to roost at about
sunset on December evenings on a slender twig of a shrub
growing in a marshy stream. The three birds pressed close

together, their yellow breasts all turned the same way, in a surprisingly exposed situation only about three feet above the water. While daylight faded, they repeated their plaintive calls. They appeared to be adults of this more social flycatcher that breeds cooperatively.

In the evening, after most other birds have gone to rest, Gray-capped Flycatchers, often in company with Vermilion-crowns, continue to hawk for flying insects in the fading light. Noisy at this time, they shout loudly and harshly before, in deepening twilight, they retire. One year they slept gregariously in a field of young sugarcane, into which they dropped after catching insects in surrounding trees while the sun sank below the hills. Their voices continued to issue from the depths of the bright green foliage of the canes, above which, now and again, one of them shot up to snatch one more insect. At least ten or twelve Gray-caps slept in this small patch of cane, where they were difficult to count. Each rested on the flat surface of a broad, arching leaf blade, at the top of the curve where it was horizontal. With the flycatchers slept many Southern Rough-winged Swallows.

Kingbirds and their close relatives, so vehemently defensive of their nests and young that they are responsible for the family name tyrant, are at other times among the most social of flycatchers. In chapter 1 I told how on both their southward and northward migrations Eastern Kingbirds roost communally. When Scissor-tailed Flycatchers arrive in the southwestern United States in spring, they pass the night in large companies. Males continue to sleep gregariously in trees throughout their sojourn in their breeding area, returning around sunrise to the nests they attend but separating from their families to join the roost in the evening. Except while incubating their eggs and brooding their nestlings, females appear to sleep in these roosts, where nonbreeding females and failed breeders also pass the night. As soon as they can fly well, young birds are led to these nocturnal gatherings. In Texas in August, as day ended Fitch counted more than two hundred Scissor-tails flying into a large Osage orange tree from all sides, singly or several together. Until the flycatchers migrated southward in early October, this roost was occupied by birds that assembled there from at least a mile around.

Scissor-tails continue their habit of roosting together after reaching their winter home, where I found them sleeping in orange trees in a small town in northwestern Costa Rica, in the noisy central park of its capital city, and among low trees in a broad marsh. In the morning they disperse in all directions, to forage singly through the day and gather again at nightfall.

The Scissor-tails that slept in the orange trees shared their roost with a smaller number of Tropical Kingbirds. In another village in northwestern Costa Rica, where both species slept amid dense foliage of fig trees, the kingbirds outnumbered the Scissor-tails. In the Guatemalan highlands in October, I found eleven kingbirds resting in a treetop in the late afternoon, but farther south I have seen no more than two or three adults keeping company in the daytime.

As the sun sank low on December evenings, scores, possibly hundreds, of Fork-tailed Flycatchers from the surrounding open country converged to roost in orange trees in a hamlet in southern Costa Rica. If I approached their trees while they were settling down for the night, they would dart out on all sides, their long tails whistling in the air. Then, after I vanished into a nearby house, they reassembled in the same orange trees. After nightfall, when they slept, I could move quietly beneath them without disturbing their repose. In the morning after they awoke, they lingered a while before they spread over the grasslands where, as told in chapter 2, they rested in low bushes to ground-gaze. With them I frequently found small flocks of wintering Yellow-rumped Warblers, who foraged around the bushes where the flycatchers rested. These two species, so different in appearance and habits, often shifted together from from one part of the grassland to another, as though drawn together by a strong attraction.

Yellow-olive Flycatcher, *Tolmomyias sulphurescens*. Sexes alike. Southern Mexico to northern Argentina.

Although most flycatchers roost amid foliage, a few prefer better shelter. All three species of phoebes, the Black, Eastern, and Say's, have been found sleeping beneath the eaves of a house, on porches, or in abandoned buildings. Through much of tropical America, Yellow-olive Flycatchers attach their pensile nests to exposed twigs and vines at the forest's edge or in a garden or pasture. Shaped much like a chemist's retort, the nests have a downward-pointing spout leading up to a rounded chamber where the eggs are laid and the young are reared. The female, which without her mate's help builds this blackish structure of fine rootlets and fibers, begins to sleep alone in it as it nears completion, often a week or ten days before she lays her first egg. After her young fly, she does not lead them back to lodge in the nest, as more careful parents with such well-enclosed structures do, but permits them to roost in the open while she continues to sleep in it. If all goes well, she may pass her nights in her hanging nest for as long as four months; but few nests remain habitable so long during the season of almost daily rains, or they are claimed by other small birds or by wasps while they are unguarded during the day. Despite her ability to build a snug bedroom for herself, the female Yellow-olive Flycatcher roosts in the open after her breeding

Retort-shaped nest of
Yellow-olive Flycatcher.

nest decays or is occupied by other tenants. The male, as far
as I can learn, roosts amid foliage at all seasons.

The Eye-ringed Flatbill takes better care of herself. Her
nest has the same retort shape as that of the Yellow-olive
Flycatcher, but it is bulkier and less neatly made, with vegetable fibers of various kinds and colors, fragments of dead
leaves, and tufts of green moss. Some of these nests are built

for eggs and young; others, which may be smaller, with shorter spouts or none, are for sleeping only—an egg would roll out of some of them. At all seasons I have found Eye-ringed Flatbills sleeping in these dormitories, always singly. Because the sexes are alike, I have not learned whether these sleepers are always females or whether males also build nests for their comfort, which I doubt, for males of this species neither help to build breeding nests nor attend the young. In Panama, I found Olivaceous Flatbills lodging in similar nests. For a few nights before she lays her first egg and after her young fly, a female Sulphur-rumped Fly-catcher may sleep in her swinging nest, but the dormitory habit is not as well developed in this bird as in the Yellow-olive Flycatcher and the flatbills.

High above treeline in the Andes, Black-billed Shrike-Flycatchers, Rufous-naped Ground-Flycatchers, and other flycatchers avoid freezing night temperatures by sleeping in caves with a diversity of other birds. Sleeping flycatchers snuggle their heads into the feathers of their shoulders, thereby protecting their eyes and diminishing loss of heat through their bills. Rarely, flycatchers increase their hours of activity by catching insects attracted to street lights or other brilliant illumination in the middle of the night, as Great Kiskadees and Tropical Kingbirds sometimes do in Brazil, and Scissor-tailed Flycatchers do in Texas.

4
Dawn Songs and Flight Songs

Flycatchers' calls, voiced to indicate their locations or to maintain contact with their mates, are so diverse that they serve, often better than plumage, to identify species. These calls may be loud or low, soft or harsh, clear or sibilant, rattles or trills, sometimes of insectlike weakness—their variety is endless. Distinctive calls have given their vernacular names to several flycatchers: phoebe, pewee, petchary (the Jamaican name for the Gray Kingbird), and kiskadee, which Venezuelans hear as *cristofué*, Brazilians as *bemtevi*, and Argentinians as *bienteveo* ("I see you well").

Although flycatchers use their voices freely, few sing well. Their most sustained recitals are their twilight songs, delivered by males chiefly as dawn songs before sunrise and less frequently in the fading light of evening. Much briefer are the greeting songs, uttered as mates unite after a separation, or duets by the pair. While flycatchers choose

nest sites, build, or attend eggs or young, they sing little dit-
ties appropriate to the occasion. Some species sing in flight.

Let us begin our survey of twilight singing with a bird
whose recital has won high praise for its conformity to the
standards of Western classical music, the Eastern Wood-
Pewee, the dawn singer most likely to be heard by many of
my readers. Around half past three in early summer in the
northern United States, the small gray bird mounts to a high
perch and repeats his simple song with variations that di-
minish monotony. One to which I listened while moon and
stars still shone brightly consisted of two phrases, *peeé-we*
and *pe-wé*, both delivered with rising inflection, with the
emphasis alternately on the prolonged first syllable and on
the shorter final note. At unpredictable intervals, the singer
varied this regular alternation with a phrase in falling inflection that often passes unnoticed by the human listener.
While singing, the bird pivoted from side to side, and often
he about-faced. For as long as forty-five minutes, the wood-
pewee may continue with hardly a pause; but around sunrise, if not earlier, he ceases; and he rarely utters his song
again until, in the dim twilight of late evening, he resumes
it for a shorter interval. This twilight song appears to be a
musical elaboration of the pewee's plaintive call note, which
he repeats frequently through the day.

The somewhat similar dawn song of the Western Wood-
Pewee, which I have not heard, appears to be less musical
than that of his eastern cousin. Among other northern dawn
singers are the Eastern Kingbird, who may begin earliest of
all, at the first hint of night's ending, the Scissor-tailed Flycatcher, the Eastern Phoebe, the Great Crested Flycatcher,
and the Least Flycatcher. The song of the last-mentioned,
which was studied by Peggy M. MacQueen, is a simple performance consisting of the more or less rhythmic repetition,
at a maximum rate of about sixty times a minute, of the
bird's characteristic call note, *che-bec*. Early in the nesting
season it may begin, after a series of preliminary calls, almost an hour before sunrise and continue for up to seventy

minutes or until a quarter of an hour after sunrise. As the season advances, the recital begins later and is less prolonged. While he sings, the Least Flycatcher moves from tree to tree along the boundaries of his territory, answering other Least Flycatchers performing in the same manner on their territories.

One great advantage of bird study in the tropics, where days are shorter than at higher latitudes in summer, is that the birds begin and end their diurnal activities at hours more convenient for many of us. We need not so drastically curtail our sleep to hear the dawn songs of tropical flycatchers, some of which are worthy accompaniments for the more gifted of the thrushes and finches who may be singing at the same early hour. Others are so quaint and rapid that they might set the tempo of a fantastic elfin dance; yet others seem forlorn dirges; a few impress us by their insistent harshness; but many have little to recommend them unless it be the tireless persistence of their repetitions.

Of the many dawn songs of flycatchers that I have heard in the tropics, for sweetness and purity of tone I assign first place to those of the big, yellow-breasted species of *Myiodynastes*. The twilight song of the migratory Sulphur-bellied Flycatcher is an utterance of rare beauty. Resting in the dim gray light of early dawn on an exposed leafless twig at the very top of a tall tree, he repeats tirelessly, in a soft, liquid voice, *tre-le-re-re, tre-le-re-re*, often continuing for more than a quarter of an hour with scarcely a pause. It is hard to believe that the bird who sings so coolly and serenely at daybreak is the same one who later in the day calls with high-pitched, strained, excited notes, often seeming to shout *weelyum, weelyum*, like an old woman screaming in a cracked voice for her distant grandchild.

The big Streaked Flycatcher, which closely resembles the Sulphur-bellied in plumage, is an equally fine singer. From a high, conspicuous station, he repeats his soft, sweet, pellucid *kawé teedly wink* almost continuously for nearly half an hour before sunrise. Like the Eastern Wood-Pewee, he

Streaked Flycatcher, *Myiodynastes maculatus*. Sexes alike. Breeds from tropical Mexico to northern Argentina, winters from Costa Rica to southern Brazil.

sometimes sings after sunset, and occasionally, more briefly and in a more subdued voice, in full daylight. One March, a partial eclipse of the sun prompted a Streaked Flycatcher to begin his crepuscular song soon after four o'clock in the afternoon. The related Golden-bellied Flycatcher of the rainy Costa Rican highlands also has a beautiful dawn song, a soft, musical *tre-le-loo,* which occasionally he repeats late in the morning, close by the mossy nest that his consort has built in a nook amid the matted roots of epiphytes high in a moss-draped tree or on a verdant cliff beside a waterfall. The high, sharp calls of this little-known flycatcher contrast with his liquid dawn song.

As befits one of the largest of the flycatchers, the Boat-bill has one of the loudest, most stirring dawn songs of all that I have heard. Perching in the gray dawn high in a tree, sometimes that in which his mate will soon build her nest, he reiterates a spirited, ringing note that sounds like *cheer.* At irregular intervals he punctuates these clear calls with a contrasting slurred note, *bo-oy.* He may continue this monologue for over half an hour. I have not heard Boat-bill mates sing in the evening twilight. A high-pitched *choip* and a whining *churr* are their usual daytime calls.

Very different is the dawn song of the Boat-bill's frequent neighbor, the Tropical Kingbird, whose voice seems incongruously thin for a bird so big and self-assertive. One of the earliest of the dawn singers, he breaks his night-long silence when the eastern sky begins to brighten. Often from a perch only a few feet above the ground, he voices his high-pitched twitter, the clear, pleasant notes rising as though by steps in two or three series, each consisting of a few rapidly trilled syllables. Usually each ascending sequence is preceded by an indefinite number of short, clear notes that sound like *pit*, but occasionally this monosyllable is omitted between two series of trills. The whole song may be represented thus:

where the short dashes represent the *pit* and the zigzag lines the trilled notes. At the height of the breeding season, this song often continues, with brief interruptions, for about half an hour. In southern Costa Rica Tropical Kingbirds sing at dawn from early February or March to mid-August, longer than any other member of the family except the Mistletoe Flycatcher.

Another soft-voiced flycatcher is the Vermilion-crowned, also a neighbor of the Boat-bill and of the Tropical Kingbird. In the gray early light, the male Vermilion-crowned tunes up by repeating over and over his plaintive call notes, *chee* and *pee-ah*. Soon he introduces a longer phrase, and the song is well begun. In Panama I heard one of these flycatchers sing *pee-ah chip chip chips-a-cheéry*. The quadrisyllable is the dawn song's distinguishing feature, repeated hundreds of times while the performance continues for twenty-five or thirty minutes, from early dawn to shortly before sunrise. It is so distinctly enunciated that it suggested the name *chip-sacheery*, which in early writings I used for this flycatcher; but that has fallen into disuse because the quadrisyllable is heard only from the southern race *(Miyozetetes similis col-ombianus)* of this wide-ranging species; the race from east-ern Costa Rica northward *(M. s. texensis)* utters at best a gar-bled version.

In appearance the Gray-capped Flycatcher differs from the Vermilion-crowned chiefly in having a white forehead in-stead of white eyebrows; but the voices of these two species, which I have often heard simultaneously, are as different as they could be. When other birds raise their sweet and liquid voices to greet the returning day, the Gray-cap awakes, calls harshly *bip, bic,* and *burr* in diverse combinations, then shouts harshly, over and over, *hic, bit of a cold; hic, bit of a cold.* At least, this is what the first Gray-cap that I heard at dawn seemed to be announcing. Other individuals have voiced slightly different phrases, but all sounded as though they suffered from sore throats. Could this be the reason why Gray-caps' dawn singing is so much briefer than that of the Vermilion-crowneds and others of their neighbors?

A prolonged, mournful whistle or whine reveals, at any time of day, the presence of the Dusky-capped Flycatcher, which ranges from the southwestern United States to north-western Argentina. The male's dawn song, delivered in the same weak, querulous voice as his call note, is compounded of a short, sharply whistled *whit,* a long-drawn, plaintive, whistled *wheeeu,* and a whistle cut short, *whee-du.* These notes are delivered in varying sequences, occasionally with the introduction of a harsh trill. At the height of the breed-ing season, this recital begins as soon as the sky starts to brighten and continues with scarcely a pause until daylight has penetrated to the depths of the forest. In a hedgerow or bushy pasture, the Dusky-cap performs while flitting from branch to branch only a few yards above the ground, but in woodland this occupant of diverse habitats perches in the treetops. I have heard this song only once in the evening twilight but often late on cloudy afternoons, well before the hour of sunset. Long, harsh trills are frequent in these after-noon performances.

For quaintness, no song that I have heard equals that of the Tufted Flycatcher, a small woodland bird that ranges through the highlands from northern Mexico to Bolivia. In a high, thin voice, a Tufted Flycatcher in Guatemala repeated insistently *de bee, de bee, de bic a de bee, de bee,* and so on. In the mountains of Costa Rica, a different race of the same species sang too rapidly for me to follow his thin notes. *Bip-bip-bip-didididup-bip-bip-bibibiseer,* he seemed to proclaim. Although I cannot claim accuracy for this rendition, if it is read as fast as human lips can move, and then allowance is made for the more rapid utterance of the bird, it should at least suggest the song's oddity.

Although we cannot tell how birds feel when they sing, I know no more effective way of conveying the quality of their songs than by revealing how they impressed me. In sharpest contrast to the Tufted Flycatcher's lighthearted, breezy greeting of the dawn, the Mistletoe Flycatcher seems to start his day with a dirge. *Yer-de-de, yer-de-de,* he whispers, again and again, then adds a faint, quavering *pe-pe-pe*—the most shrinking and melancholy of all the dawn songs that I have heard. One Mistletoe Flycatcher gave variety to his recital by adding a weak little rattle or trill. The call of this diminutive bird is a low, plaintive whistle, of the same sad character as his dawn song.

Very thin and high-pitched, as befits a bird only three and a half inches long, are the Southern Beardless Flycatcher's not unmusical notes, which often reveal its presence in the crowns of leafy trees. Here in the valley of El General in southern Costa Rica this is consistently the earliest flycatcher to nest and to sing. As the dry season begins in early January, and exceptionally during a few rainless days in November, the male mounts a high, exposed twig, often at the top of a dead or dying tree, to repeat over and over in the dim early light, *te be be be,* or at times *te be be be be.* While voicing these penetrating notes, he slightly raises his crown feathers, and turns his head from side to side. One morning when the monologue was exceptionally prolonged, I counted twenty-one phrases per minute at the outset, seventeen somewhat later, but only six per minute toward the end, when the flycatcher appeared to be tiring and interjected more call notes into his dawn song. By March or early April, when neighboring flycatchers are most vocal at dawn, the little beardless flycatcher has ceased to sing and to nest.

Some dawn songs are noteworthy chiefly for their monotonous simplicity. In the eastern foothills of the Ecuadorian Andes, from late August to past the middle of October, the songs of Olive-chested Flycatchers filled the air, for no more forceful singer raised his voice to drown their humble recitals. Perched six feet up in a thicket and about forty feet from his mate's nest, one continued for twenty minutes to repeat a low and soft but full and far-carrying monosyllable, *chite,* at the rate of once every second, or sixty times per minute—well over a thousand times before he fell silent in the growing daylight. From the distance came the voices of others of his kind.

Almost equally uninspired is the dawn singing of the Mountain Elaenia in the rainy Costa Rican highlands. Perching in scattered pasture trees, often on the topmost bough of the highest of them, males began in the first gray light to call, in an odd, dry voice, *d'weet d'weet d'weet,* continuing this monotonous chant for many minutes with hardly a pause. Sometimes the elaenia introduced a little variety by singing *d'weet d'weet d'weet a d'weet,* or *d'weet d'weet d'weet d'weeger d'weet,* or even *cheet a cheet, cheet a cheet, cheet a cheet.* Begun in April, this bizarre dawn singing continued on a reduced scale through most of June, when nesting ended.

Some may object that the rapid repetition of call notes, however rhythmic, does not qualify as song and that many of these twilight recitals should be designated dawn calling. However, by their length and continuity, they resemble the songs of more melodious birds, and their biological function, although not wholly clear, appears to be that of the songs of such birds as thrushes and wood warblers.

Whereas some male flycatchers seem to awake in a

merry mood, and others appear to be dejected, still others sound angry. Before dawn, while stars still twinkle brightly overhead and only a paleness above the eastern horizon presages the approach of a new day, or even by moonlight, a Yellow-bellied Elaenia reveals his presence by a prolonged, harsh *wheer*. After several repetitions, this drowsy note is followed by a hard, assertive *we do*, which will be reiterated with hardly a pause for many minutes. Although the elaenia has a fairly varied vocabulary, his other notes are reserved for later in the day. In his twilight singing he limits himself to these two notes, *we do*, and the only latitude he allows himself is to pronounce them from time to time in a tone more vehemently insistent. *We do, we do, we do,* WE DO, *we do, we do* he calls interminably, until the drowsy listener feels forced to concede that they *do*, whatever it may be that they claim to accomplish. The impression that the singing Yellow-bellied Elaenia is angrily self-assertive is intensified by his flattened crest, like the laid-back ears of a resentful horse.

Although the closely similar Lesser Elaenia is actually a more quarrelsome bird than the Yellow-bellied, his dawn song sounds droll rather than contentious. At daybreak in the breeding season, from the low trees where these small gray birds nest and roost, issues a thin flow of quaint notes, *a we d' de de, a we d' de de, a we d' de dee*, which continues for from a quarter-hour to half an hour, until silenced by the brightening sky. After sunrise the elaenias may repeat this refrain from time to time, but rarely, unless perhaps agitated by a rival, do they recite it for many minutes on end, as at dawn.

In the evening twilight, the Lesser Elaenia does not deliver a prolonged song, as at daybreak. But in the gathering dusk, after the day's last songs have been sung and all diurnal birds are sinking into drowsy silence, he suddenly shoots above his tangled thicket, rising up and up on a steep, irregularly twisted course until he has reached or even passed the level of the highest treetops of the tropics, and as he ascends he repeats without pause his unmistakable *a we d' de de.* At the summit of his wild flight, the dim little figure promptly turns and shoots recklessly earthward, continuing his dry ditty until lost to view amid the bushes where he sleeps. On an evening in mid-April, I had the good fortune to witness the start of an ascent. While daylight was fading fast, a pair of elaenias, resting amid low herbage, repeated a curious low whistle that barely escaped harshness, turning from side to side as they called. Then, when the sky had faded to a dim gray, one of them rose into the air in a tortuous course, uttering the familiar refrain as he ascended. Turning, he darted to the ground near the spot whence he had risen, and both elaenias fell silent for the night. Frequent in the evening twilight, the Lesser Elaenia's flight song is rarely heard at dawn.

The flight songs of flycatchers are less widespread, or have been less often recorded, than their dawn songs. Throughout its vast range from the southwestern United States to central Argentina, the male Vermilion Flycatcher rises high in the air to repeat songs that appear to vary with the individual, the locality, and the occasion. Near the southern extremity of its distribution, Hudson heard the flycatcher repeat "his remarkable little song, composed of a succession of sweetly modulated metallic trills," while "the bird mounts upward from thirty to forty yards, and, with wings very much raised and rapidly vibrating, rises and drops almost perpendicularly half a yard's space five or six times, appearing to keep time to his notes in these motions." In Mexico, William Beebe heard the Vermilion Flycatcher sing in flight *ching-tink-a-le-tink.* Early on a January afternoon, while I wandered among steep, bushy slopes in the Cauca Valley of Colombia, one of these flycatchers rose up to circle, singing, above the tops of the low trees. His display incited other males to perform in the same fashion, and these in turn stimulated others more distant to do the same. At times three of these brilliant birds were simultaneously in the air, circling slowly and singing a simple verse of two notes, much like the dawn song that I had heard in Ecuador.

PLATE 1
Ochre-bellied Flycatcher, *Mionectes oleagineus*. Sexes alike (5 in, 12.5 cm).
Southern Mexico to western Ecuador, Bolivia, and Amazonian Brazil.

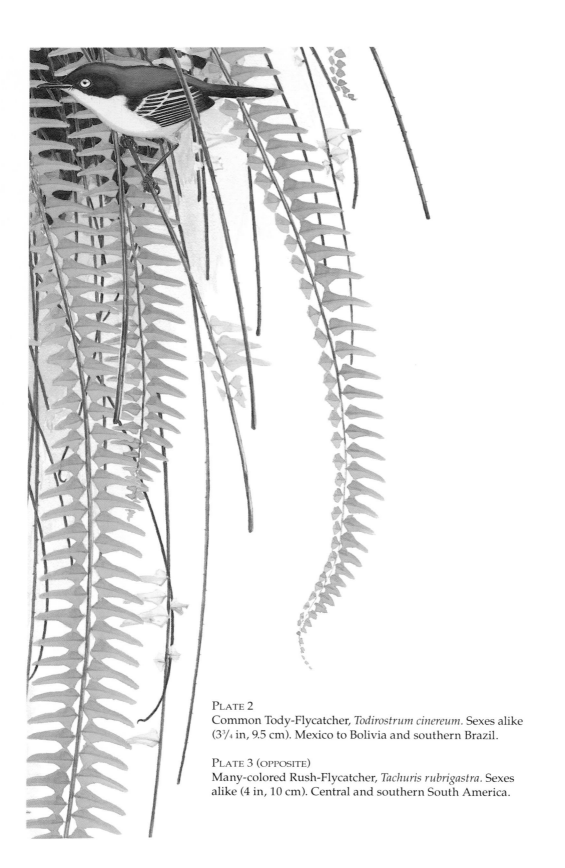

Plate 2
Common Tody-Flycatcher, *Todirostrum cinereum*. Sexes alike
(3³/₄ in, 9.5 cm). Mexico to Bolivia and southern Brazil.

Plate 3 (opposite)
Many-colored Rush-Flycatcher, *Tachuris rubrigastra*. Sexes
alike (4 in, 10 cm). Central and southern South America.

PLATE 4 (OPPOSITE)
Golden-crowned Spadebill, *Platyrinchus coronatus*. Sexes alike (3¼ in, 8.5 cm). Honduras to western Ecuador, Amazonian Brazil, and the Guianas.

PLATE 5
Royal Flycatcher, *Onychorhynchus coronatus*. Detail of head with crest expanded. Sexes similar (6¾ in, 17 cm). Southern Mexico to north-western Peru, northern Bolivia, and southern Brazil.

PLATE 6 (OPPOSITE)
Sulphur-rumped Flycatcher,
Myiobius sulphureipygius. Sexes
alike (4³/₄ in, 12 cm). Southern
Mexico to western Ecuador.

PLATE 7
Ornate Flycatcher, *Myiotriccus
ornatus*. Sexes alike (4¹/₂ in, 11.5
cm). Andes from Colombia to
southeastern Peru.

Plate 8 (opposite)
Yellow-bellied Flycatcher, *Empidonax flaviventris.* Sexes alike (4³/₄ in, 12 cm). Breeds from northern Canada to northern United States, winters from southern Mexico to eastern Panama.

Plate 9
Vermilion Flycatcher, *Pyrocephalus rubinus. Left,* female (5¹/₂ in, 14 cm), *right,* male (same size). Southwestern United States to Nicaragua; Colombia to Guyana and central Argentina, Galápagos Islands; southern breeders migrate to eastern Colombia, eastern Ecuador, and eastern Peru.

PLATE 10 (OPPOSITE)
Crowned Chat-Flycatcher, *Ochthoeca frontalis*. Sexes alike (5 in, 13 cm). Eastern Andes of Colombia to northern Bolivia.

PLATE 11
Bright-rumped Attila, *Attila spadiceus*. Sexes alike (7 in, 18 cm). Northwestern Mexico to western Ecuador, Bolivia, and southeastern Brazil.

PLATE 12
Great Crested Flycatcher, *Myiarchus crinitus*. Sexes alike (8 in, 20 cm). Breeds in eastern and central Canada and United States, winters from southern Florida and Cuba to northern South America.

PLATE 13
Western Kingbird, *Tyrannus verticalis*. Sexes alike (8 in, 20 cm). Breeds from southwestern Canada to northwestern Mexico, winters from southern Mexico to Costa Rica.

PLATE 14
Scissor-tailed Flycatcher, *Tyrannus forficatus*. Sexes similar (7³/₄ in, 19.5 cm to tips of central rectrices). Breeds in south-central United States and northeastern Mexico, winters from southern United States to western Panama.

PLATE 15
Sulphur-bellied Flycatcher, *Myiodynastes luteiventris*. Sexes alike (8 in, 20 cm). Breeds from southeastern Arizona to Costa Rica, winters east of Andes in Peru and Bolivia.

PLATE 16
Great Kiskadee, *Pitangus sulphuratus*.
Sexes alike (9 in, 23 cm). Northwestern Mexico and southern Texas to
central Argentina.

Fork-tailed Flycatcher, *Tyrannus savana*. Sexes similar, female shorter tailed. Breeds from southeastern Mexico to central Argentina, winters from southeastern Mexico to northern Argentina.

Black Phoebe, *Sayornis nigricans.* Sexes alike. Southwestern United States to northwestern Argentina.

Hudson also described how, just before sunset, groups of Fork-tailed Flycatchers (which he called Scissor-tail Tyrants) rise to the treetops, calling to one another with loud, excited chirps, then shoot upward like rockets to a great height in the air, where they whirl around for a few moments before hurtling downward with the greatest violence, opening and closing their wings in their wild zigzag descent and uttering a succession of sharp, grinding notes. After this wild performance, they separate in pairs that perch in the treetops and join in uttering castanetlike rattles.

In Costa Rica, early on a September morning, I watched a Black Phoebe circling slowly in the air with body almost vertical while he repeated *fe-be, fe-be* over and over. This flight song hardly differed from the plaintive dawn chant that the phoebe continues monotonously for many minutes from a high stance, sometimes a rooftop. Eastern Phoebes also fly in circles in the air, repeating their usual call.

Among northern flycatchers, the Least ascends high into the evening air to circle above the treetops, rising and falling, while he repeats *che-bec, che-bec,* much as in his dawn song. In the western United States a related species, the Gray Flycatcher, shortly before sunset, hops excitedly to the top of a tree, from which he flutters out in a wide circle, singing, *whit-whit-whit-whit-whit-wheak-wheat-wheat-stíddle-d-doo-stíddle-doo-stíddle-d-do*. After finishing this jumble of bubbling notes, the small bird dives vertically into the tree-top. Ned K. Johnson also heard another western *Empidonax*, the Dusky Flycatcher, singing in similar fashion, not only at sunset but more frequently around midday and in the late afternoon.

Willow Flycatchers, in the same genus, have ways of singing peculiar to themselves. In northwestern Montana, Winton Weydemeyer found they contributed little to the many-voiced chorus that greets the new day, often not be-ginning to repeat their simple songs until after sunrise. Through most of the day, they sing more persistently than most neighboring birds, even through the warm hours of midday and early afternoon when many birds are silent. A rest of two or three hours in late afternoon prepares them for their grand evening chorus. Starting soon after sunset, more and more individuals are heard until, in the darkness fol-lowing twilight, suddenly all the flycatchers sing together in a swelling chorus that lasts, with brief intermissions, for about twenty minutes. Then, in the gloom, it suddenly ends, long after all other diurnal birds have become silent. Weydemeyer only occasionally noticed Willow Flycatchers rising high in the air to sing.

In southern Wisconsin, Willow Flycatchers behaved dif-ferently. Here Robert McCabe found flight songs frequent. After an accelerated series of *wheet-wheet* calls, a flycatcher changes to *creeet-fitz-bew* in increasing tempo as he leaves his perch and spirals skyward to a height of thirty to fifty feet above the marsh where he dwells. At the zenith of his ascent he ceases singing and dives abruptly down to his original song perch or one nearby. The same individual may rise, singing, from two to, rarely, six times in an evening. Only about a quarter of the birds that participate in the twi-light chorus fly up singing on any one evening. Starting after sunset, these songful ascents continue for as much as half an hour after civil twilight, which terminates when the sun falls six degrees below the horizon and it is too dark for most outdoor occupations.

The biological significance of twilight and flight songs is not always clear. Instead of performing along the edges of their territories, in the manner of the Least Flycatcher, nearly all the flycatchers I have heard sing at dawn stayed too continuously in one place to be demarcating their boundaries—if indeed their foraging areas had definite limits. Only exceptionally have I found a bird singing in its nest tree, and often no dawn song was audible near a nest attended by a male of a dawn-singing species. Nevertheless, the fact that a territorial dispute evokes dawn songs at hours when they are not ordinarily heard suggests that they are not unrelated to territory. If this is true, why do these flycatchers not proclaim their domains more continuously through the day, as many birds in other families do?

Since dawn singing is practiced by flycatchers that have long had mates and are already nesting, it is not, as a rule, for the purpose of attracting partners. Probably, if dawn songs are more than an outlet for energy stored up during a restful night, while waiting for daylight to become bright enough for effective foraging, their chief function is to inform neighbors of the same species of the singer's location, so that they may avoid nesting too near. Often, while listening to one bird sing at dawn, I have heard no other of the same kind. Similarly, the much rarer flight songs reveal where birds are settled, and may help to prevent too close clustering by birds that are not colonial. But these flights, too, may be primarily an expression of exuberance, comparable to the high soaring on thermal updrafts by wide-winged birds bigger than flycatchers, apparently only for the exhilaration of such ascents.

5

Duets, Greetings, and Nest Songs

Although flycatchers are not the most vocally gifted of birds, many have varying notes for different occasions, and they are far from silent. Mates often duet. Hudson described how Great Kiskadees (his Bienteveo Tyrants), upon rejoining their partners after a long separation, perch close together, their yellow bosoms almost touching, and with crests raised and wings beating the branch "scream their loudest notes in concert—a confused jubilant noise that rings through the whole plantation. Their joy at meeting is patent, and their action corresponds to the warm embrace of a loving human couple" (I:178).

The greetings of the smaller but similarly attired Vermilion-crowned Flycatcher are more restrained, as befits their less ebullient temperament. As one member of a pair alights beside the other, they flutter their wings and together voice high-pitched notes that are almost a trill. While perching quietly not far apart, from time to time they repeat this soft, chiming utterance, which serves not only as a salutation but likewise as an assurance that each is within hearing of its partner and all is well with them. When, after approaching with shallow, mincing wing-beats, a Tropical Kingbird alights beside its consort, both spread and vibrate their wings while they twitter together. A pair of Common Tody-Flycatchers, foraging through the foliage of a tree, keep in touch with clear, resonant little trills, uttered simultaneously or in sequence. This responsive trilling is not neglected while the female incubates her eggs or broods her nestlings; often from her swinging nest she trills in response to her partner. Like many permanently resident flycatchers, these live in pairs throughout the year and trill to each other at all seasons. While Yellow Flycatchers circulate in pairs through thickets and bushy pastures, they maintain contact by an almost continuous pleasant chatter.

When after a brief separation members of a pair of Torrent Flycatchers join each other on a boulder rising above a rushing current, they face each other and, lifting up their heads, sing in unison a short refrain composed of sharp *chip* notes rapidly repeated. This simple twittering lacks musical quality, but the notes ring out above the incessant clamor of the torrent. Similarly, when mated Sooty Flycatchers (which Hudson more poetically called Little Riverside Grey Tyrants) meet, they erect their crests and flick their wings while together they voice trills and hurried sharp notes. Likewise, the related White-crested Flycatchers (not to be confused with White-crested Elaenia) sing together "a little confused song". Hudson attributed to the Tufted Tit-Flycatcher (his Little Tit-like Grey Tyrant) "a little shrill duet." As mated Scissor-tailed Flycatchers approach each another, they flutter their wings, fan their tails, and twitter with staccato notes. Least Flycatchers and Hammond's Flycatchers chatter and twitter when they meet at or away from their nests. Such greetings, widespread in the family, appear to be expressions of an emotional temperament.

The strangest of all the duets I have heard is that of the Yellow-bellied Elaenia. It was not at first evident to me that this utterance came from more than a single throat, yet it seemed incredible that any bird could produce notes so intricately garbled. Enlightenment came one day when I stood between the members of a pair. One elaenia gave, as a signal, a single harsh whistle of a peculiar quality. Its mate repeated a slightly more mellow note a number of times, with an undulatory effect. The elaenia who had given the signal for this performance joined in with similar notes, but their duet was harsh and confused because they failed to keep time. If the partners are close together when they unite their voices in this fashion, the listener might never suspect that he or she heard more than a single bird trying to sing with garbled notes.

Nest songs are usually low twitters or rapid flows of soft notes, often heard while flycatchers seek nest sites, build, incubate, and attend their young. While they choose a place for their nest, a pair of Gray-capped Flycatchers seem to confer together. Probably the female actually makes the choice while her mate attends her closely, taking great interest in all she does. Sitting in a prospective nest site, the Gray-cap utters a long-drawn soft rattle or twittering *churr*. Similar sequences of subdued notes are given by Vermilion-crowned Flycatchers as they examine sites for their nests. The nest song of one flycatcher stimulates others to murmur in low voices. Early one afternoon, while pairs of Gray-caps and Vermilion-crowns were singing in this fashion as they tested sites in neighboring orange trees, a male Boat-billed Flycatcher sat in the nest his mate had just started and softly twittered as she perched beside him, holding a stick for the structure. While a pair of Golden-bellied Flycatchers investigated nooks and recesses in epiphyte-burdened trees where they might build their nest, one of them, whose sex I could not learn, repeatedly pushed into promising hollows amid the epiphytes' matted roots, to the accompaniment of low twitters. While a female of the related Sulphur-bellied Flycatcher filled an old woodpecker hole with sticks, her mate from time to time rested in front of the doorway, twittering in a subdued, slightly harsh voice. While a female Tropical Pewee busily built, she often voiced a little, clear, trilled *cheee*.

Many female flycatchers murmur their nest songs more or less frequently while they incubate. Sitting in her domed nest where I could see her through the wide doorway in the side, a Gray-cap repeated a quaint, low, cackling song throughout the day, but most often in the early morning and late afternoon, giving the impression that she enjoyed sitting peacefully upon her three speckled eggs, with her mate nearby. At intervals she shouted sharply and loudly, answering his equally loud calls—which seemed imprudent behavior that might draw hostile attention to her nest. Gray-caps are at all times such voluble birds that it must be difficult for them to restrain their voices, even while attending their nests. At each of two nests of Vermilion-crowned Flycatchers that I watched, I heard low nest songs as the females, returning from their recesses, passed through their doorways and snuggled down on their eggs. Thereafter, while incubating, one of these Vermilion-crowns was silent, but the other often sang in an undertone. Sitting in nests stolen from these two yellow-breasted species, female Piratic Flycatchers murmur rhythmically or warble softly notes pleasant to hear and unexpected in birds that begin their nesting with so much rude clamor.

Nest songs are sung by a wide range of incubating flycatchers from big Boat-bills to tiny Golden-crowned Spadebills. Opening her mouth widely to reveal a blackish lining, the sitting spadebill emits a weak, high-pitched, insectlike buzz. Incubating in her pensile nest, a diminutive Black-headed Tody-Flycatcher continually repeated a slight, soft trill. From a Dusky-capped Flycatcher covering three eggs in a hollow bamboo stem, from time to time I heard a rapid sequence of low, soft notes that seemed a lullaby. Slightly different utterances, which might be called feeding songs, are heard from parents offering food to nestlings, especially to those no more than a few days old when they are slow to open their mouths or to swallow what is placed there.

Critical points in a nesting sequence—selecting the nest site, hatching, and the nestlings' departure—often evoke nest songs. In a cluster of liverworts hanging beneath a calabash tree, a female Mistletoe Flycatcher had built a mossy globe with a side entrance and had laid two dull white, speckled eggs. One April afternoon, while I set up a ladder to look into this nest, she clung at the doorway, head inside, uttering notes so low that I would not have heard them if I had not been so near. After continuing this soft murmur for about half a minute, she entered the nest and settled down to brood, head outward. Apparently, she had been so absorbed by what was happening in the nest that she had failed to notice me climb the ladder, for ordinarily Mistletoe Flycatchers are more wary. Presently, becoming aware of her visitor, she darted out and protested with the low, mournful notes usual on such occasions. Looking into the chamber, I found a nestling in the act of hatching, a little of its tiny pink body visible between the separating parts of the shell. Its mother had been so strongly moved by the sight that she forgot her customary caution, as few birds do.

A Lesser Elaenia reacted strangely to the hatching of her nestlings. While incubating, she had frequently repeated a typical subdued nest song; but while she sat in her nest in the dim early light of an April morning, covering a newborn nestling and a hatching egg, she voiced a simplified version of the male's dawn song. Since I had never heard an undoubted female perform in this manner, I attributed the notes to her mate in a nearby tree until, watching the nest with my binocular, I saw her mandibles move with each repetition of the dry *de weet*. While she sat there, he brought food for the nestling. She continued this chant for nearly half an hour before sunrise.

Early one morning while its parents were absent, a fledgling Vermilion-crowned Flycatcher flew from its nest and alighted in a neighboring tree. When a parent, apparently the mother, returned with a billful of food and for the first time found the young bird in the open, she began a nest song. The rapid flow of low, sweet notes continued for minutes while she rested amid the sun-drenched foliage of a guava tree, her fledgling beside her. A parent Gray-capped Flycatcher also sang profusely when it joined a fledgling who had just flown from the nest. These yellow-breasted flycatchers begin and end their nesting with pleasant little songs that appear to express strong parental feeling. Male flycatchers, who are not known to incubate, deliver such songs less frequently, occasionally voicing them when they come to inspect a nest with eggs or to feed their nestlings.

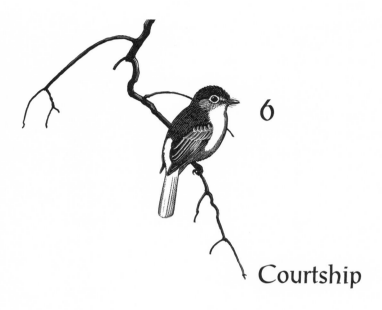

6

Courtship

In the tropics, where a large proportion of the permanently resident birds live in pairs and maintain their territories throughout the year, courtship, the formation of pairs, and the establishment of territories are not such prominent activities as they are among the migratory birds of higher latitudes. With more time at their disposal, tropical birds arrange these matters more leisurely and obscurely, with less singing and skirmishing than do northern birds with short breeding seasons. Not only the flycatchers that live in dense thickets and heavy forests where they are difficult to keep in view but even those of more open country win their nuptial partners in ways that are not obvious, and often long before the breeding season.

Among tropical flycatchers, I have rarely seen what I interpreted as the first stage in pair formation. In South America Lesser Elaenias migrate in large numbers, and in southern Costa Rica most of them disappear after July or August, although where they go is unknown. In late January, familiar notes drew my attention to the first of these small gray flycatchers that I had seen in months. In bright morning sunshine, two of them were flitting around in a guava tree. One, apparently the male, was repeating a short, dry phrase that sounded much like the elaenia's dawn song, but with the phrases more widely spaced. The other answered each repetition of the droll *a we d' de de* with a more liquid monosyllable, *weer*. After each change of position, they alighted from a few inches to a few feet apart. Even when they moved from tree to tree, they kept close together. After this play had continued for many minutes, the elaenia that had been answering *weer* changed her tune, replying to the first with a subdued version of his dry *a we d' de de*. They continued to flit and to call for more than half an hour before they flew over the crest of the hill and were no longer seen or heard. It is significant that this courtship, if such it was, was by birds that had just arrived on their breeding ground after a long absence.

While I watched a Royal Flycatcher's nest where the female was incubating, she returned to her eggs escorted by her partner. Alighting on a vine near the nest, he spread his scarlet crest widely while turning his head from side to side so vigorously that the splendid tiara quivered. Simultaneously, he shook his half-opened wings, fanned out his yellowish tail, and rapidly uttered peculiar notes, somewhat like his usual piping call. This superb display, the most elaborate that I have seen in many hours of watching Royal Flycatchers, lasted only a few seconds. At its conclusion, apparently unimpressed, the female entered her pensile nest. This appeared to be a repetition of the display by which this Royal Flycatcher had already won her.

Aerial displays are prominent in the courtship of the Eastern Kingbird and its relatives. Rising with shallow, quivering wing-beats to a height of perhaps fifteen or twenty feet above a tree, the Eastern Kingbird flies a zigzag course of alternate ascents and short dips, repeating this exhibition again and again to the accompaniment of shrill cries. Occasionally the dip becomes a long, slow dive. The Scissortailed Flycatcher, essentially a kingbird with greatly elongated outer tail feathers (as is recognized by the latest classification), performs a similar courtship flight. After mounting with rattling wings to perhaps a hundred feet in the air, he rises and falls with sharp reversals, tracing a course that may be represented by a string of Vs, at the angles rapidly spreading and closing his long, black-and-white streamers. While engaging in this aerial seesaw, he screams loudly or emits a rolling, cackling sound like high-pitched hand clapping. At the apex of his last upward flight, which may take him still higher, he topples over, making two or three reverse somersaults that display the soft orange beneath his wings. Cassin's Kingbird also performs "crazy zigzag sky dances" while calling loudly.

As told in chapter 4, Fork-tailed Flycatchers in Argentina precipitate themselves downward from their social evening ascents in wild zigzag courses. In Brazil, however, they descend in a series of tight spirals, with wings spread like parachutes and tails widely fanned. This flight, illustrated by Helmut Sick, appears to be a courtship display. In contrast to the solo displays of other kingbirds, a male and a female Gray Kingbird rise into the air together, either straight upward or in a spiral, crossing each other as they ascend and twittering loudly, as described long ago by Audubon and confirmed by subsequent observers.

Very different were the antics of Golden-crowned Spadebills that I watched in the lower levels of rain forest. At intervals, beating its wings deeply, one of these tiny birds flew jerkily between perches, while in the air making a peculiar sound not unlike that produced by twanging a tightly stretched rubber band but louder. It appeared to be made by the bird's wings, in the manner of certain manakins. The spadebill's flights during which I heard the peculiar sound ranged from one or two to about ten yards in length, between branches fifteen to thirty feet above the ground. When short, such flights might be sharply inclined but when long they were nearly horizontal. While flying and while resting between flights, the diminutive bird tirelessly repeated a prolonged vocal trill or buzz. One morning when two spadebills were in view, these performances continued for about a quarter of an hour. The sexes of spadebills are alike and I was not sure that all the displays were by the same individual, but the two never performed simultaneously. The bird who flew and twanged appeared to be courting the other who silently watched. Spadebills live in pairs.

The courtship of a number of small migrant flycatchers appears to be largely a matter of chasing a prospective partner and following her around. The sexes of wood-pewees and many small species of *Empidonax*, such as the Gray Flycatcher, the Alder, and the Western, are so similar in appearance that a male may be able to learn the sex of an intruder into his zealously defended territory only by its behavior. Accordingly, he chases every visitor. Another male usually flees beyond his boundaries, or more rarely stays to fight for possession. A female does neither but stubbornly stays within the territory, perhaps avoiding his first angry onslaught by taking refuge amid sheltering foliage, thereby revealing her sex and readiness to become his partner. This mode of disclosing sex by passive resistance to territorial males is employed by females of other families, such as the Song Sparrow and the European Robin.

A minority of flycatchers never form pairs; the males remain aloof from nests that females attend alone. Best-known of these nonpairing species is the very plain, greenish olive Ochre-bellied Flycatcher, widespread in humid lowland forests from Mexico to Bolivia. Throughout a long breeding season, a male is to be found daily in a small area, about thirty to one hundred feet in diameter, in mature forest or older second-growth woods. He does not confine his display to a preferred perch but moves between slender, bare, horizontal branches of saplings or small trees, usually from ten to thirty feet above the ground, rarely as high as fifty feet, on each of which he pauses to repeat his tuneless song. A variable number of well-spaced, low, weak notes precede a more rapid series of fuller notes in measured time. *Whip wit whip wit wit chip chip chip chip chip chip*, the little bird chants, scarcely interrupting his monologue while he flits about and plucks insects from foliage. While performing, he slightly raises the feathers of his unadorned crown and throws up his wings momentarily above his back, one at a time and usually alternately, a gesture that helps to draw attention to his obscure self in the dim light of the underwood. Females flip up their wings in the same manner, but in silence. Often several male Ochre-bellied Flycatchers associate loosely in a dispersed courtship assembly or lek, each within hearing of at least his nearest neighbors. Just what happens when a female visits these displaying males is difficult to learn because the sexes differ little in appearance and visibility in the underwood is so limited. After insemination, the female goes to a distance to lay her eggs in her wonderful nest.

Although Sulphur-rumped Flycatchers never form cooperating pairs, I have noticed no display other than the drooped wings, spread tail, and exposed bright yellow rump of both sexes while they forage through the forest. Rarely, from birds that were apparently males, I have heard utterances elaborate enough to be called songs, one of which consisted of five sharp notes followed immediately by about the same number of warbled notes. In wet foothill forests from Costa Rica to Peru, male Scaly-crested Pygmy-Flycatchers, perching from about fifteen to thirty feet above the ground, call attention to themselves by repeating sharp metallic notes, singly or up to six or eight rapidly reiterated. They do not call from a special perch, but each is to be found day after day in the same small area. Calling males are too widely scattered through the woodland to form a courtship assembly. Their helmetlike crests of elongated cinnamon-rufous feathers with black centers are usually worn flat, like the

Ruddy-tailed Flycatcher,
Terenotriccus erythrurus.
Sexes alike. Southeastern
Mexico to northern Bolivia,
central Brazil, and the Guianas.

more colorful crest of the Royal Flycatcher. These male pygmy-flycatchers call too persistently through the day for months together to participate in nesting; at the single nest I could find and watch, only the female fed the young.

Likewise, female Northern Bentbills attend their nests alone, while the males call with low growls, croaks, and throaty churrs amid lush vegetation at the forest's edge or in dense thickets. Male Ruddy-tailed Flycatchers twitch up both short wings together above their backs while perching, and on short flights from twig to twig they vibrate their wings to make a whirring sound much as certain manakins do. They also remain aloof from the females' nests. The solitary females of these four species all build pensile structures entered from the side or below. None of these males emancipated from parental chores has become much more colorful than the females of their species, as the males of many hummingbirds, manakins, birds of paradise, and others with similar reproductive habits have done.

Several flycatchers of southern South America have curious displays, the significance of which is not clear from available accounts; but I suspect that, like the more northern birds just profiled, they fail to form cooperating pairs. The females of these austral birds, however, differ greatly in

Spectacled Flycatcher,
Hymenops perspicillatus.
Male. Brazil and Bolivia
to Argentina and Chile.

plumage from the males, which is unusual in the flycatcher family. The male Spectacled Flycatcher (Hudson's Silverbill) is black, with the primaries white except at the base and tip. His bill and the conspicuous ruffled wattle around each eye are primrose yellow, appearing silvery white at a distance. In the marshes and open grassy places where he dwells, he perches upright at the top of a tall stalk or bush, from which he shoots vertically upward as though propelled by a steel spring, making a humming sound with his wings, which in flight flash out white that is usually hidden. At the zenith of his ascent, perhaps fifteen yards up, he emits a little shrill cry while he turns a somersault before dropping back to his perch. He is said to chase violently the dark brown females of his kind, probably in nuptial pursuits.

Another small bird with a similar display flight is the Cock-tailed Flycatcher, whose peculiar tail was described in chapter 1. With rapidly beating wings and looking more like a butterfly than a bird, the male rises slowly to a height of five to ten yards, then drops with an insectlike *tic-tic-tic*. Another small black bird with white on its wings, Hudson's Black-Flycatcher, has a different display. Through much of the day the male perches on the exposed summit of a bush, from time to time darting out to seize a passing insect. Suddenly, he quits his perch to fly in circles around it as swiftly as a moth gyrating around a candle flame, uttering sharp, clicking notes while his wings hum loudly, their white bands looking like a pale mist enveloping the bird.

Gifts of food help some birds to win their mates. Typically the male feeds the female, but sometimes this is reversed. In flycatchers, feeding appears not to occur in the early stages of pair formation; only for the Vermilion Flycatcher and the Eastern Phoebe have I found reports of a male feeding his partner before she has begun to incubate. During incubation, feeding of the female by the male is more widespread, or has been more often seen and reported; but this is nuptial rather than courtship feeding and will be considered in chapter 8.

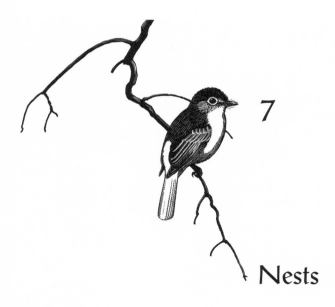

7

Nests

Often I have watched pairs of the large yellow-breasted fly-catchers—Vermilion-crowned, Gray-capped, Golden-bellied, Streaked, and Tropical Kingbird—seeking nest sites together. One member of the couple sits in an attractive spot, trying out how it feels; the other rests close beside it, while they voice the low, twittering nest songs described in chapter 5. The day after a pair of Tropical Pewees sat side by side in a bare crotch of a spiny tree, the female started to build in this same fork. In other species the male selects sites and tries to persuade his consort to build in one of them by nest-site-showing displays. Displays by the Vermilion Flycatcher have been described by several writers, most vividly by Hudson, who told how, a few days after the arrival of this migratory bird in Argentina, the male visits the chosen spot about once a minute, "sits on it with his splendid crest elevated, tail spread out, and wings incessantly fluttering, while he pours out a continuous stream of silvery gurgling notes, so low they can scarcely be heard twenty paces off." Occasionally the partners sit briefly side by side in a nest site. Male Pied Water-Flycatchers, Acadian Flycatchers, and Eastern Phoebes also try to draw their mates' attention to promising sites, one of which females may accept.

Flycatchers build a greater diversity of nests than does any other family of New World birds except the ovenbirds. Most of their nests are open cups that differ greatly in construction. The broad, shallow nests of Tropical Kingbirds are made of slender herbaceous vines, rootlets, tendrils, fine woody twigs, grasses, and similar coarse, dry vegetable materials, with finer pieces in the lining. Often they contain so little of these materials that the eggs are visible through the mesh of the sides and bottom. Lengths of vines may dangle untidily far below the supporting branch. Neglecting opportunities for concealment, these kingbirds frequently build their nests in exposed situation, low or high, where they enjoy a wide outlook and can spy and boldly confront enemies before any come near. Nevertheless, their nests are often plundered, probably by night, when the valiant birds' vigilance does not avail.

Among the great diversity of situations chosen by Eastern Kingbirds for their bulkier nests are frequently boughs projecting above the margin of a river or pond, sometimes no more than a yard or two above the water, or even stubs rising from the water. Scissor-tailed Flycatchers' open cups are roughly built but with more soft materials than the kingbirds use, often much cotton, sheep's wool, feathers, rags, twine, scraps of paper and, where available, Spanish moss (*Tillandsia*). The bulky bowls of Boat-billed Flycatchers are loosely constructed of materials that the female laboriously pulls from trees instead of gathering them from the ground: fibrous roots of epiphytes, rhizomes of ferns that creep over branches, small living orchid plants, pieces of dead vines, and crooked twiglets, all forming a fabric so open that light sometimes passes through the bottom. Boatbills' nests are situated far out on leafy branches of trees standing isolated in clearings, usually from twenty to one hundred feet above the ground.

As a rule, the smaller the bird, the softer the materials it selects and the neater its nest. That of the Eastern Wood-Pewee is a dainty cup of plant fibers, strips of bark, and other flexible pieces, so well covered on the outside with lichens that it blends with the horizontal branch or fork to which it is firmly attached, often in a site with little screening foliage. Some wood-pewees' nests contain green leaves, which are rarely used by flycatchers. The nest of the tropical Yellow-bellied Elaenia resembles that of the northern pewee. The soft but firm walls of the shallow cup are composed of fine rootlets, plant fibers, bits of herbaceous stems, and other vegetable fragments, all closely matted and bound together with generous applications of cobweb, a material so indispensable to small avian builders that if they ate all the spiders they might become extinct because they could not build their nests and reproduce. The same sticky strands attach gray and green foliaceous lichens to the exterior of the elaenia's nest, so thickly that the structure matches the lichen-covered branch that supports it. The interior is well padded with fine fibers and many downy feathers, often from domestic chickens, which stand up conspicuously above the rim.

In the interior of rain forests, Golden-crowned Spade-bills find other materials to fashion nests as soft and neat as those of hummingbirds. Situated in upright forks of shrubs or saplings, the beautiful little cups are composed of leaf skeletons, large fern scales of chestnut color, light-colored bast fibers, richly branched stems of mosses and liverworts from which the minute leaves have fallen, and fungal rhizomorphs that creep over decaying stems and are called vegetable horsehair. The outside is plastered with few or many shreds of white spiders' egg cases, usually not enough to conceal the nest's generally brownish color.

The nest of the Many-colored Rush-Flycatcher is built

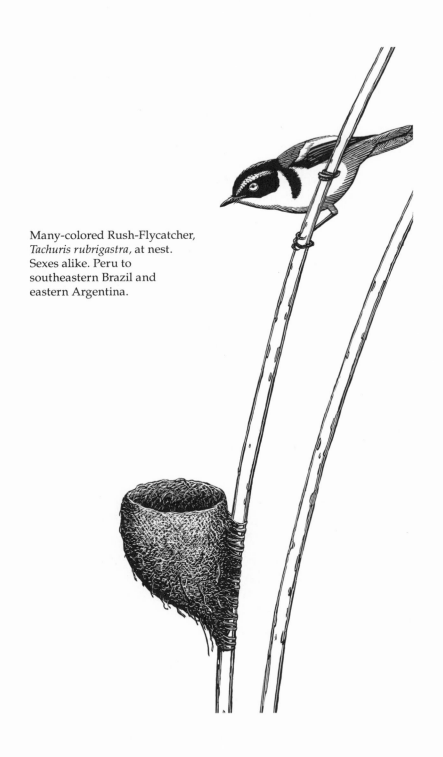

Many-colored Rush-Flycatcher,
Tachuris rubrigastra, at nest.
Sexes alike. Peru to
southeastern Brazil and
eastern Argentina.

with such consummate skill that it has won the highest admiration of all who have seen it. Shaped like an inverted cone about six inches high, it is attached to a single polished rush about midway up and two or three feet above the water of a South American marsh. The soft fragments of yellow sedge of which it is made are so well cemented together with some gummy substance that the surface is perfectly smooth, as though the nest had been shaped in a mold. The cup's circular rim is somewhat constricted, the better to retain the eggs when a breeze sways the rushes.

Most cup-shaped nests are placed in trees, shrubs, or other plants that support them from below. A few, including those of the Acadian Flycatcher and the Bran-colored Flycatcher, hang, vireo-fashion, between the arms of a horizontal fork to which the rim is fastened. Cinnamon Flycatchers build cups of mosses and lichens on small ledges or projections of vertical cliffs, where they are exposed to view but difficult to reach. The Cliff Flycatcher chooses for its nest site a rocky ledge, window sill, air conditioner, or other flat surface of a building, on which it lays a circle of pebbles to outline the position of its open cup of twigs and rootlets cemented together with saliva, like nests of swifts. Black Phoebes and Eastern Phoebes attach massive brackets, made of pellets of mud bound together with fragments of plants, to almost any vertical surface sheltered from rain, such as the face of a cliff or overhanging boulder, the piers and stringers beneath bridges, the wooden walls of sheds or empty houses, even the sides of wells yards below ground level. The hollow in the top is lined with fine grasses, feathers, or other soft materials; the outside, at least of the Eastern Phoebe's nest, is covered with green moss.

A few flycatchers build open nests on the ground: Rufous-backed Negritos in marshy places in Chile and Argentina; Chocolate Flycatchers on stony soil with sparse vegetation in wind-swept Patagonia; Strange-tailed Flycatchers on the grassy pampas of Argentina. Yellow-bellied Flycatchers hide their mossy nests on or near the ground, often amid sphagnum moss in northern coniferous woods. On barren rocky slopes of the southern Andes and in neighboring treeless country, ground-flycatchers (*Muscisaxicola*) and some shrike-flycatchers (*Agriornis*) build their cupped nests deep in clefts in cliffs, in caves, or amid piles of stones.

Open nests, or at least the neater of them, are fashioned by procedures widespread among passerines and hummingbirds. To shape the interior, the builder presses down her breast and with rapid backward kicks of both feet pushes it against the wall, often turning to face in different directions, making it round. Her tail rises almost straight up into the air while with her breast she smoothes the cup's bottom. With her bill she arranges the materials, chiefly on the outside, pulling loose ends upward and inward, tucking them into the rim. Thus, the wall is built up and made compact by pushing and pulling its components in opposite directions by feet and bill. Its height often appears to be regulated by the base of the builder's tail. Pieces too long, stiff, or thick to fit well in the fabric, or that hang untidily below it, may be carried away. Later, lichens or silken egg cases of spiders may be attached to the outside. By such procedures the elegant nests of elaenias, wood-pewees, and spadebills are shaped and embellished.

More elaborate than the cups are domed structures with an entrance in the side. These roofed nests, supported from below in trees or shrubs, low or high, are made of straws, weed stems, seed down, or almost any pliable material available to the builder. Bulky and externally ragged, they are smooth and softly lined within. The roof, thick enough to shed rain, often extends forward to shield the entrance. Such structures are built by Gray-capped and Vermilion-crowned flycatchers, other species of *Myiozetetes*, and Great Kiskadees. Occasionally, a Gray-cap or a Vermilion-crown, finding a recently abandoned open cup of some other bird, converts it into a closed structure by roofing it over, suggesting how these domed nests may have evolved. Of these covered structures the bulkiest are those of the Great

Kiskadee, with thick walls that often include, among other components, whole nests of much smaller birds, such as seedeaters. Untidy balls of dry vegetation with a sideward-facing entrance are also built by Pied Water-Flycatchers and White-headed Marsh-Flycatchers in shrubs or trees, low above the water or rarely as high as thirty feet, and frequently in exposed situations.

Intermediate between the bulky domed nests and the freely hanging structures that will next receive our attention are the little ovoid purses with a narrow mouth in the side made by some very small birds, including the Mistletoe Flycatcher and the beardless flycatchers. With thick walls composed of diverse pliable materials bound together with cobweb, and softly lined with silky seed down, these neat, cozy pouches are supported all around, in a tuft of moss, liverworts, or lichens draping beneath the branch of a tree, in a dangling epiphytic orchid plant, amid a cluster of spiny pods of an annatto bush, in a richly branched panicle of a tall tree, or hidden in a large curled dead leaf, often one of the cecropia tree, caught up in vines high above the ground.

Pendent nests are built in great variety by small and middle-sized flycatchers. Probably most familiar are those of the Common Tody-Flycatcher, which from tropical Mexico to Peru and Brazil dangle from garden shrubbery and ornamental trees or high at the forest's edge. I have even seen them fastened to telegraph wires beside railroads. From a slender support, the nest hangs down for nearly a foot. At about the middle of this slender, elongate structure, where it is thickest, is a rounded chamber, entered through a narrow doorway shielded by a visorlike projection from its upper edge. Below the chamber loose fibers hang in a "tail" that may be long or short, or sometimes lacking. Strong fibers, often from decaying leaf sheaths of bananas or related plants, are the mainstay of these constructions. Entangled among them a great diversity of dry vegetable materials— scraps of papery bark, seed down, fine grass blades, small withered flowers—all bound together with much cobweb,

Nest of Common Tody-
Flycatcher, *Todirostrum
cinereum*. Sexes alike. Mexico
to northwestern Peru and
southern Brazil.

make the walls thick and firm. These nests often appear
ragged, but they admirably protect eggs and slowly devel-
oping nestlings from hot tropical sunshine and drenching
downpours. Similar nests are built by other species of
Todirostrum.

Nest of
Sulphur-rumped
Flycatcher,
*Myiobius
sulphureipygius.*

Quite different is the pensile nest of the Sulphur-
rumped Flycatcher, fastened to a dangling epiphyte root, a
freely hanging vine, or the tip of a slender leafy twig, in an
open space well below the forest canopy or above a shady
watercourse. From a slender point of attachment, the nest

expands broadly downward to enclose a rounded chamber with a rather wide doorway that is shielded and concealed by a curtain continuous with the outer walls. To enter her nest, the flycatcher flies straight upward and passes through an enclosed space, or antechamber, to reach her eggs or nestlings. The walls of these nests are made of brownish fibrous materials, and the bottom of the chamber, where the eggs rest, is thickly padded with light-colored fibers. Black-tailed Flycatchers build similar nests.

Also similar but more elongate were nests of Ruddy-tailed Flycatchers that I found in the understory of a Panamanian forest. Pyriform in shape, they tapered from a rounded bottom to a pointed apex, where they were attached to a thin drooping twig, a dangling vine, or to a palm frond. The entrance to the chamber was protected by a projection from the wall above it. Composed of dark fibrous roots, fine brown fibers, fragments of dead leaves, bits of decaying sticks, small pieces of bark, and other shreds of vegetable debris, these nests were blackish.

Nests of the Sepia-capped Flycatcher and the related Slaty-capped Flycatcher hang low in dark places, beneath an overhanging streamside or roadside bank, or under a huge rock or log, where they are attached to a short length of root or tendril. The only Slaty-capped's nest that I have seen hung from a splinter beneath a great shattered trunk that projected over the rapids of a boulder-strewn stream, deep in foothill forest. Between globular and pyriform in shape, with a round doorway shielded by a visorlike projection that seemed superfluous in this sheltered situation, the nest was only four feet above the rushing water. About eight inches long, it had thick walls composed almost wholly of fine rootlets, with a small admixture of light-colored fibers. The ample globular egg chamber was lined all around with pale, finely shredded bast fibers and some tufts of silky seed down. In it rested a single nestling in pinfeathers, who cried shrilly when I illuminated its nursery with an electric bulb attached by a cord to my flashlight and inserted a small mirror to look in.

The female Northern Bentbill hangs her small pyriform nest only a foot or two above the ground, amid sheltering vegetation at the forest's edge, beside a shady rural lane, or in a weedy coffee plantation. The structure is composed almost wholly of pale bast fibers, with an incomplete covering of green moss. The narrow round doorway in its side gives access to a softly padded chamber where two eggs or nestlings lie. The Black-capped Pygmy-Flycatcher's nest, similar in shape, is well covered with green moss and hangs higher, from four to twenty-two feet up in lowland rain forest or in a forest clearing. The Yellow-olive Flycatcher's retort-shaped blackish pensile structure, which serves as a dormitory as well as a receptacle for eggs and young, and the bulkier, less tidy nest of the Eye-ringed Flatbill are described in chapter 3.

The structures that Royal Flycatchers suspend above shady streams are among the most extraordinary of birds' nests. From the point of attachment at the end of a slender drooping branchlet or dangling vine, the female extends her structure both upward, around the supporting branch or vine, and downward below its tip. She gathers long, brown, threadlike female inflorescences of *Myriocarpa*, a small tree of the nettle family frequent along watercourses, and wraps them around the branch. Among these tangled strands she sticks skeins of cobweb, pieces of dead leaves, coarse roots of epiphytic orchids, slender rhizomes of epiphytic ferns, palm fibers, and a little green moss, continuing to add pieces until her nest is two to nearly six feet long. Near the middle of this slender mass, always where it hangs free of its support, she opens a shallow niche for her eggs and nestlings. This alcove occupies such a small part of the elongated nest that it is not always easy to find. My first Royal Flycatcher's nest, swinging above a mountain torrent in Honduras, was so different from any bird's nest that I had seen that I did not know what it was until the flycatchers' interest in it prompted me to investigate. The curious accumulation of vegetable fibers hung above my head as I bal-

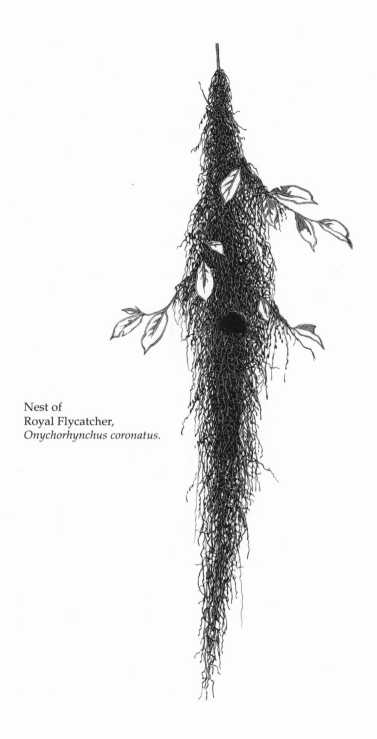

Nest of
Royal Flycatcher,
Onychorhynchus coronatus.

anced precariously on a ledge of rock, and I could not find the niche with the eggs until the birds showed me exactly where it was.

The most beautiful of these pensile structures is built by the plainest of their builders, the Ochre-bellied Flycatcher, who attaches it to a vertically hanging root of an epiphytic aroid beside the massive trunk of a forest tree or beside a cliff verdant with mosses, ferns, and other small plants. The pyriform nest may be three or four feet long if the loose ends of dry *Myriocarpa* inflorescences are included in the measurement, but the massive body of the structure occupies somewhat less than half of this length. It is wholly covered with moss that except in very dry weather remains green. The middle layer consists of pieces of leaf skeletons, light-colored bast fibers, and other strands. The chamber is amply lined with fibers, often from the decaying center of a palm trunk. The triple wall of mossy covering, fibrous middle layer, and soft lining is about half an inch thick. Above the chamber, the neck of the pear-shaped structure is solid and composed largely of moss. The entrance is a narrow orifice in the side. The Olive-striped Flycatcher builds a similar nest.

Unable to weave, flycatchers build their pensile nests in a manner very different from that of the orioles and oropendolas, who skillfully lengthen their pouches from the top downward as neatly woven fabrics. The flycatcher begins by accumulating a solid mass of tangled fibers and other materials. When this grows large enough, the builder pushes into it, forcing the materials apart, making a small pocket in the sides or bottom. As the hollow expands with further pushing, the flycatcher strengthens the walls by carrying more pieces inside. She also attaches them to the exterior by alighting at the top and inserting them into the mass with a deft movement of her bill as she creeps rapidly downward over the outside of the growing nest. With a quick revolution of her whole body, she winds more strands around the top of her nest and its support to strengthen its attachment.

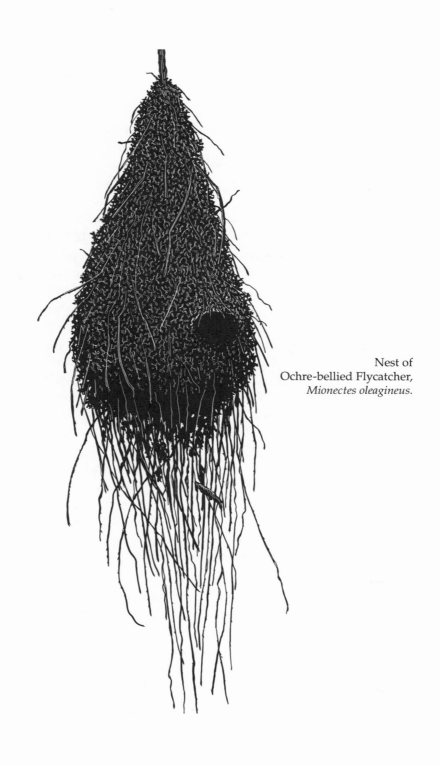

Nest of
Ochre-bellied Flycatcher,
Mionectes oleagineus.

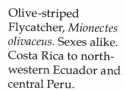

Olive-striped
Flycatcher, *Mionectes
olivaceus*. Sexes alike.
Costa Rica to north-
western Ecuador and
central Peru.

Gradually the structure takes its final shape while the walls thicken. A flycatcher's hanging nest is not woven but matted.

A small minority of flycatchers nest in closed spaces that they cannot prepare for themselves. The Great Crested Flycatcher and other species of *Myiarchus*, the Sulphur-bellied Flycatcher and other species of *Myiodynastes*, the Long-tailed Flycatcher, White-ringed Flycatcher, and White Monjita build cupped nests in old woodpeckers' holes and other cavities in trees, or in bird boxes and such miscellaneous crannies as they can find. Rufous-tailed Flatbills and other species of the South American genus *Ramphotrigon* lay their eggs in hollow broken-off ends of rotting stubs, a yard or less above the ground in woodland. The cavity often opens upward, affording no protection from rain. Great Crested Flycatchers and related species frequently add strips of shed snakeskin to their nests, possibly, as some naturalists have surmised, as a talisman to hold nest-plundering snakes aloof, but more probably because these exuviae are soft and flexible, like shreds of cellophane or plastic that are also found in birds' nests.

When a Sulphur-bellied Flycatcher chooses a deep cavity for nesting, she laboriously fills it with coarse sticks to within a few inches of the doorway, and on top of this accumulation she fashions with petioles and thin sticks a cup for her eggs. She prefers to sit where she can see (and be seen) through the entrance, instead of hiding out of sight, in the manner of many hole-nesting birds. In addition to old woodpeckers' holes, Cattle Flycatchers rear their young in a room of the Rufous-fronted Thornbirds' many-chambered nest of interlaced sticks, in an abandoned clay nest of the Rufous Hornero, or in a Firewood-gatherers' great castle of sticks. This commodious structure is coveted as a nest site by other birds; but the doughty Cattle Flycatcher can hold it against most competitors, even a flock of six or eight Bay-winged Cowbirds, the only species of *Molothrus* that is not a nest parasite.

Unique in the family are the habits of the Piratic Flycatcher. Not known to build a nest for itself, it steals newly made, closed nests of other birds, including the domed structures of the Great Kiskadee and of the Vermilion-crowned Flycatcher and other species of *Myiozetetes*; the pensile, retort-shaped nests of the Eye-ringed Flatbill and of the Yellow-olive Flycatcher and other species of *Tolmomyias*; the bulky structures of White-winged Becards; the long, swinging pouches of oropendolas, caciques, and orioles; even the chamber that Violaceous Trogons have carved for their eggs amid the brood combs of a great papery wasps' nest hanging high in a tree. The Pirates' strategy is simple but effective. While the chosen victims build their nest, the indolent thieves perch nearby, watching, waiting, and calling with seemingly insolent levity. After the coveted nest has been finished and contains an egg or two, the Pirates, becoming more aggressive, incite the builders to chase them. While one member of the outrageous pair tolls the nest's owners away, the other slips into the unguarded structure and throws out an egg. This stratagem is repeated until the nest is empty, whereupon the owners abandon it and start a new one. In a gesture of nest building, the Pirates carry in a few dry leaves, lay their eggs upon the loose litter, and rear their families as competently as though they had built the roof that covers them. Their persecution may cause Yellow-rumped Caciques to abandon a small colony.

As one would expect, female Ochre-bellied Flycatchers, Sulphur-rumped Flycatchers, bentbills, and others that do not form pairs build their nests alone. Even monogamous males that valiantly defend the nest and diligently feed the young rarely help to build. Among the few species in which males are known to contribute substantially to nest construction are the Common and Slate-headed tody-flycatchers, Yellow-bellied Elaenia, and Torrent Flycatcher. Male Pied Water-Flycatchers bring materials to the nest, deposit them there, and leave to their mates all the work of shaping the structure.

Males of other species pick up nest materials only to toy with them and drop them; or, following their building partners, they may carry pieces to the nest without depositing them there. A male Boat-billed Flycatcher brought the same rootlet thrice to the fig tree where his mate built and thrice carried it away when she flew off for more material and he followed. Once, resting near the nest that she was shaping, he held a rootlet for eleven minutes, then dropped it to chase away a trespassing bird. He might have saved his mate much strenuous labor, tearing rootlets from epiphytes, if it had occurred to him to pass the piece to her, or to her to take it from him. Mostly, males do no more than escort their toiling consorts back and forth on their excursions to gather materials. Or a male may rest near the nest, lazily preening, and greeting his mate with low notes and fluttering wings as she passes by on her trips to and fro. Often males remain for long intervals beyond view.

The time required to build a nest varies with the season and the complexity of the structure. Nests started early, long before the birds will lay eggs in them, usually take longer than those begun at the height of the breeding season. A Lesser Elaenia finished her open cup in about four days. Open nests usually take a week or ten days, and domed nests may be completed almost as promptly. Pensile nests, the most complex of all, are rarely built in less than two weeks and may take much longer. Common Tody-Flycatchers devote as much as five weeks to the construction of their hanging nests, and a Yellow-breasted Flycatcher in Suriname spent about the same time on her retort-shaped structure. Flycatchers taking so long to finish a nest build at a leisurely pace that they can accelerate when necessary. The pair of tody-flycatchers that dawdled so many weeks over their first nest of the season completed a replacement nest in only ten days. This operation was accelerated by salvaging materials from the earlier nest, which had been emptied by a predator. Many birds that will not lay again in a nest that has been violated tear it apart, using its materials to build a new one.

The rate of building is also highly variable. When working fastest, a flycatcher may add a billful of material to her nest about once every two minutes. A Lesser Elaenia brought pieces to her cup twenty-seven times in an hour. A Royal Flycatcher made sixty-three trips to her nest in two hours and a Southern Beardless Flycatcher sixty trips in the same interval. Hudson's statement that a building Vermilion Flycatcher "visits her nest a hundred times an hour with invisible webs in her bill" may be slightly exaggerated. The highest rates of building are rarely long sustained and may be followed by intervals of little or no activity.

A Yellow-bellied Elaenia laid her first egg the day after she finished her nest, which was exceptional. Much more often the completed nest stands empty for two to ten days and occasionally longer, especially in tropical forests. I have recorded intervals of nine and eleven days in the Golden-crowned Spadebill and seven and thirteen days in the Royal Flycatcher. Many tropical birds breed on a more leisurely schedule than would be practicable at higher latitudes where the favorable season is much shorter.

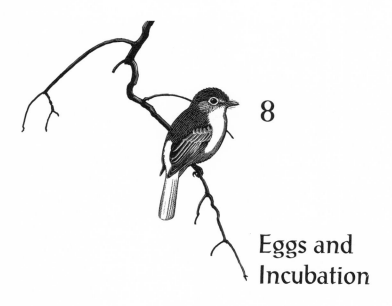

8

Eggs and Incubation

In the inner tropics, within about twelve degrees of the equator, most flycatchers regularly lay only two eggs; some species lay two or three; but sets of four are rare. Flycatchers that breed at higher latitudes, both north and south, lay from two to six eggs, the most frequent number probably being four. Even in the same wide-ranging species, such as the Tropical Kingbird and the Black Phoebe, the number of eggs increases with distance from the equator. In other families, too, the size of broods increases with latitude, a phenomenon that has been attributed to the long summer days, which give parent birds more hours for foraging for themselves and their young; but the greater mortality of birds that either perform long, hazardous migrations or try to survive through harsh winters must be compensated by a higher rate of reproduction. Among tropical flycatchers, two is the most frequent number of eggs whether both parents attend the nest or only one, although two parents should be able to feed twice as many young as single parents can. Apparently, factors other than ability to nourish the brood influence its size.

In the tropics, flycatchers' eggs are usually laid on alternate days, less often on consecutive days, rarely with an interval of three days. Even in the same set, the interval may vary; two days between the first and second eggs, for example, and only one between the second and third, or in the reverse order. In the North Temperate Zone, where the breeding schedule is abbreviated, laying on consecutive days appears to be more frequent than it is in the tropics. Western Flycatchers skipped no more than one day while laying a set of four or five eggs; in some sets of four, all were laid on consecutive days. Scissor-tailed Flycatchers deposited three to five eggs without missing a day. Unlike many tanagers and finches, tropical flycatchers rarely deposit their eggs before sunrise. Many of their eggs are laid about the middle of the forenoon, some nearer midday.

Flycatchers' eggs are white, pale gray, buff, or cream, sometimes immaculate but more often more or less heavily spotted or flecked with rufous, brown, chocolate, or pale lilac, with the markings concentrated in a wreath around the thicker end and scattered elsewhere over the surface. Among the most beautiful eggs are those of the kingbirds, the Sulphur-bellied Flycatcher, and the Streaked Flycatcher, which are glossy white or cream decorated with rich shades of cherry, reddish brown, and lavender. The heavy pigmentation of eggs of the Sulphur-bellied Flycatcher and other species that breed in cavities suggests that this is a character retained from ancestors that laid in open nests, where darker colors and marks that disrupt the eggs' outline make them less conspicuous. Unlike these secondary hole-nesters, primary hole-nesters such as kingfishers and parrots, with ancestors that for long ages have nested in closed spaces, lay plain white eggs. The exceptionally dark ground color of the Piratic Flycatchers' eggs hints that these birds have acquired the habit of stealing covered nests from other birds only recently, in evolutionary time.

Darkest of all the flycatchers' eggs that I have seen are those of the Royal Flycatcher. Almost solidly deep reddish brown on the thicker end, they pale to dull buff on the opposite pole. The coloration of these eggs is as unusual as the form of the nest in which they lie, and both have apparently evolved together. Their dark shade makes them less conspicuous in a shallow open niche in the midst of the greatly elongated nest.

Because the sexes of most species of flycatchers are so similar in appearance, to learn their participation in incubation has not been easy. Sometimes close watching of a pair will reveal slight differences between a male and female that to casual observation are exactly alike; the feathers of the male's crown may be slightly tousled, or a broken feather may serve for temporary identification. Certain early accounts affirmed that male flycatchers share incubation. Long ago, while still doubtful about this, I found a Yellow-bellied Elaenia's nest with a single egg and fastened above it a tiny brush dipped in white paint. While trying to remove the foreign object, one member of the pair acquired a white spot on its forehead. The next day, this marked bird laid an egg while I watched, thereby revealing her sex. Continued watching showed that only this individual sat in the nest. In subsequent studies of this and many other tropical flycatchers, I have never found a male helping to incubate. Others have watched both tropical and temperate-zone flycatchers without gathering evidence that males incubate, with the single exception of the cooperatively breeding White-bearded Flycatcher.

Yellow-olive Flycatchers, which use their nests as dormitories, begin to sleep in them when they near completion, often a week to ten days before they deposit the first egg, and they continue this habit throughout the period of laying. Sulphur-rumped Flycatchers may start to sleep in their nests as much as four days before the first egg appears.

Other flycatchers who do not use their nests as dormitories may begin to pass the night in the nests while they contain only the first egg, or before the set is complete. They are also found sitting by day on incomplete sets. Such premature covering of eggs leads to sequential rather than simultaneous hatching; but the interval between the hatching of the first egg and that of the last may be less than the time that the earlier egg or eggs were prematurely covered because the parent did not then apply full heat to them.

Most flycatchers with roofed nests sit in them with the head outward, looking through the doorway; but Sulphur-rumped and Royal flycatchers regularly incubate and brood with the head inward, tail projecting through the entrance. The Sulphur-rump's dark tail is screened by the apron over the nest's opening; the Royal Flycatcher's tawny-orange tail sticks far outward, where in a spot of sunshine filtering through the foliage above it might be mistaken for a colorful freshly fallen leaf caught up among the tangled mass of materials that constitute the nest. It is so much brighter than the olive-brown back of the bird, sitting in the open niche with her head turned sideward, that it does not seem to belong to her.

While sitting in their nests, incubating or brooding, flycatchers are often far from silent. From time to time sitting birds repeat the twittering nest songs described in chapter 5, as though contented to have the smooth-shelled eggs in contact with the bare brood patch that warms them into life. Now and again mates exchange calls or sing duets. Irrepressibly voluble Gray-capped Flycatchers shout out loudly from their domed nests, revealing their locations. At intervals the incubating flycatcher rises up to adjust her eggs with her bill, sometimes spreading her crown feathers to expose the bright patch in their midst when she bends down her head. Sometimes, usually in species that line their nests with downy feathers, a flycatcher returning to her nest after a recess from incubation brings something to add to the nest.

After accompanying an incubating partner while she

Black-and-white Tody-Flycatcher,
Todirostrum capitale. Sexes alike.
Southeastern Colombia to
northeastern Peru and
southwestern Brazil.

forages, the male flycatcher often escorts her back to the nest. As a female Common Tody-Flycatcher flies toward the round doorway in the side of her pensile nest, her mate often follows so closely that he appears to be trying to reach it first; but she always wins the race and enters, while he veers aside, then utters a little dry rattle which she answers from within. If she prolongs her absence, he flies toward the doorway without her, trying to induce her to return. Similarly, a male Bright-rumped Attila races with his mate to the nest in a nook between the buttresses of a great tree, or amid massed epiphytes, but he too always permits her to arrive first, while he continues onward. Such spectacular escorting is widespread among birds whose nests have sideward-facing entrances; male euphonias and chlorophonias, in the tanager family, regularly fly with their consorts to nests in which they neither incubate nor brood.

Male flycatchers may spend much time near the nest, keeping in contact with their sitting partners by voice, or they may remain out of sight and hearing for long intervals. Some of them tend to guard the nest while the female is away foraging; but of those I have watched, only the big Boat-bill guards somewhat consistently from a nearby perch. Although he may not remain near the nest, the male Tropical Kingbird, surveying his domain from a high lookout, is alert to descry and repel approaching enemies. At intervals, while his mate is absent, a male flycatcher perches on the rim of an open nest, or in the doorway of a covered one, and looks intently in while murmuring softly. Rarely, he enters briefly. Standing on the nest's rim, a male Torrent Flycatcher gently preened the plumage of his incubating consort. In contrast to such affectionate intimacy, peppery Lesser Elaenias and Yellowish Flycatchers repulse mates that come too near the nest while they incubate, not permitting their approach until after the eggs hatch and fathers are needed to help feed the nestlings.

Only exceptionally do male flycatchers feed their incubating partners. Among the few species for which nuptial feeding has been reported are the Great Crested, Tufted, Scrub, Vermilion, Least, and Western flycatchers; Eastern Wood-Pewee; Tropical Pewee; and Eastern Phoebe. One male Tropical Pewee fed his mate on or near the nest twelve times in four and a half hours; another, ten times in four hours. David E. Davis saw a male Least Flycatcher feed the incubating female eleven times in six hours at one nest, twelve times in two hours at another. In other species, nuptial feeding is less frequent or absent. While his mate was at recess, a male Tropical Pewee came occasionally with a small insect in his bill and lowered his head into the cup, as though trying to deliver it to the unhatched eggs. A male Tufted Flycatcher similarly offered food to eggs. Less often than males, female birds bring food before their young are born. While her eggs were still a week from hatching, a female Dusky-capped Flycatcher entered her nest with an insect in her bill and crooned softly to them. Likewise in other families of birds, parents impatient to feed their young occasionally anticipate their hatching, leading one to wonder whether they have a mental image of their unborn progeny.

In fair weather, small flycatchers incubate restlessly, alternating brief sessions with absences of about equal length. They may pass only half the time, or a little more, on their eggs. An insect flying near the nest may tempt a sitting bird to dart out and overtake it, after which she may return promptly or continue to forage. Larger flycatchers sit more steadily, often devoting about three-quarters of the daytime to incubating. Mistletoe Flycatchers and the elaenias, which eat many berries, tend to incubate more constantly than do species of similar size that depend almost wholly on insects for their nourishment. Table 1 summarizes records that I made while watching flycatchers of twenty-seven species incubate their eggs, mostly in rainless weather. Each line gives the record for a single nest, watched for five to twelve hours, sometimes throughout a day but more often, for the longer records, during the morning of one day and the afternoon of another. Records made at nests of two individual birds of the same species, sometimes with intervals of years, are often closely similar, suggesting that the rhythm of incubation is an innate character of the species, only slightly modified by external conditions that are not extreme.

The constancy of incubation, given in the column at the far right, is the percentage of a bird's active day that it spends in its nest. When the nest has been watched from dawn to dusk of a single day, this is simply the total of all the sessions divided by the time elapsed between the incubating bird's first morning departure from her nest and her final return for the night, multiplied by 100. When the record is for less than a whole day, constancy (T) may be calculated by computing the average length of the sessions (S), the average length of the recesses (R), then deriving the percentage of time on the nest by the formula

$$T = \frac{100S}{S+R}$$

Because of the different ways that incubation data for North American flycatchers have been presented, they do not fit into table 1. With automatic devices, S. Charles Kendeigh recorded the attentive behavior of an incubating Great Crested Flycatcher during five full days. Her 25 daily sessions averaged 21.3 minutes in length, her intervening

TABLE 1   *Incubation Patterns of Female Tropical Flycatchers*

Species	Weight (grams)[a]	Hours watched	Sessions (min.) Num-ber[b]	Range	Aver-age	Recesses Range	Aver-age	Constancy %
Common Tody-Flycatcher	6.3	8	19	3–19	8.8	3–49	16.6	35.0
Slate-headed Tody-Flycatcher	7.5	7 1/2	11	13–38	20.0	11–29	20.3	50.0
Slate-headed Tody-Flycatcher	7.5	6	11	14–33	21.0	9–16	12.7	62.0
Yellow Flycatcher	8.0	7	9	12–40	27.8	13–42	21.5	56.0
Torrent Flycatcher	8.0	5	25	3–13	6.7	1–13	5.2	56.0
Torrent Flycatcher	8.0	4 1/2	17	1–14	7.2	2–17	9.1	44.0
Mistletoe Flycatcher	8.5	6	8	22–35	27.4	10–24	13.3	67.0
Mistletoe Flycatcher	8.5	5	7	8–75	32.1	10–18	12.5	72.0
Tufted Flycatcher	8.5	5	43	0.5–12	4.4	0.5–6	2.6	63.0
Golden-crowned Spadebill	9.0	5	16	3–23	10.6	2–19	8.3	56.0
Golden-crowned Spadebill	9.0	5	10	8–27	16.3	9–20	12.3	57.0
Bran-colored Flycatcher	9.5	6	28	3–14	6.6	2–16	6.5	50.0
Yellowish Flycatcher	12.0	6	12	9–38	20.0	5–15	9.0	68.0
Sulphur-rumped Flycatcher	12.0	11 1/2	22	7–33	14.1	8–28	15.6	47.5
Sulphur-rumped Flycatcher	12.0	6	7	15–30	23.1	14–47	24.6	48.5
Tropical Pewee	12.5	4	12	3–29	12.3	2–14	6.3	66.0
Greenish Elaenia	13.0	12	22	7–49	19.4	4–37	12.7	60.5
Yellow-olive Flycatcher	14.5	7	14	10–28	17.4	7–28	13.5	56.0
Yellow-olive Flycatcher	14.5	5 1/4	9	10–36	19.7	8–52	18.1	52.0
Long-tailed Flycatcher	15.0	5	30	1–8	4.5	2–13	5.5	44.5
Lesser Elaenia	17.5	12	42	2–60	13.3	1–7	3.3	80.0
Dusky-capped Flycatcher	20.0	10	13	5–68	24.0	13–41	22.7	51.0
Mountain Elaenia	20.0	7	14	9–62	25.1	1–9	5.2	83.0
Royal Flycatcher	21.0	12	19	4–52	22.4	3–33	13.6	62.0
Royal Flycatcher	21.0	6	10	9–32	17.9	4–18	12.0	60.0
Yellow-bellied Elaenia	25.0	10	27	4–49	15.6	4–13	6.9	69.0
Yellow-bellied Elaenia	25.0	7	19	8–22	13.2	4–12	8.2	62.0
Piratic Flycatcher	26.0	6 1/4	8	20–49	34.0	5–18	12.9	72.5
Piratic Flycatcher	26.0	8	13	13–40	26.0	6–22	11.3	70.0
Vermilion-crowned Flycatcher	27.0	10	12	9–72	30.4	11–31	18.5	62.0
Vermilion-crowned Flycatcher	27.0	9	25	4–32	12.8	4–27	10.2	56.0
Gray-capped Flycatcher	30.0	12	22	7–42	20.0	7–20	11.8	63.0
Gray-capped Flycatcher	30.0	6	15	6–50	15.5	4–14	8.0	66.0
Tropical Kingbird	40.0	7	5	19–92	50.8	15–47	24.6	67.0
Tropical Kingbird	40.0	8	7	10–56	32.4	9–20	12.1	73.0
Bright-rumped Attila	40.0	11 1/2	5	63–111	93.2	13–60	38.3	67.0
Bright-rumped Attila	40.0	10 1/2	5	33–184	90.4	13–50	34.8	71.5
Streaked Flycatcher	45.0	12 1/2	13	15–72	30.3	8–18	12.6	71.0
Sulphur-bellied Flycatcher	45.0	7	15	5–32	17.0	1–21	8.5	67.0
Boat-billed Flycatcher	70.0	10	16	15–77	30.2	2–19	8.3	78.0
Boat-billed Flycatcher	70.0	6	5	27–68	44.6	5–20	13.3	77.0

[a]Approximate weights are from Stiles, Skutch, and Gardner, 1989.
[b]This column gives the number of sessions; the number of recesses may differ by 1 or 2.

recesses 13 minutes, and her constancy 61.2 percent, which is about normal for a bird weighing approximately 35 grams. During a five-day recording of incubation by the 18–gram Eastern Phoebe, the number of sessions fluctuated from 61 to 124 daily, with an average of 93.6; their average length was 5.7 minutes. Her recesses averaged 4.5 minutes, and she sat with a constancy of 56 percent. The 11–day record for the 14–gram Eastern Wood-Pewee reveals more steadfast incubation. With a mean of 32.1 sessions per day, averaging 19.6 minutes and separated by recesses averaging 6.8 minutes, she achieved a constancy of 73.5 percent. At three nests of the 11–gram Western Flycatcher, watched for 87, 70, and 94 hours by John Davis and his associates, incubation constancy was, respectively, 78, 81, and 77 percent. Incubating Hammond's Flycatchers were on their nests for 77 percent of the 19.3 hours that David E. Davis observed them. He also found Least Flycatchers incubating with a constancy of 77 percent. At seven Acadian Flycatchers' nests studied by Russell Mumford, constancy averaged 73 percent. A Yellow-bellied Flycatcher, watched by Lawrence Walkinshaw and C. J. Henry in Michigan, incubated for 81 percent of nearly 12 hours. At five Dusky Flycatchers' nests studied with self-recording instruments by Martin Morton and María Pereyra throughout the incubation period, constancy averaged 76 percent, with extremes of 72 and 82 percent.

Although during their periods of diurnal activity these northern flycatchers covered their eggs more constantly than did the tropical flycatchers of about the same size that I watched, the total time spent on their nests every twenty-four hours by these two categories of birds may not differ as greatly as the foregoing figures suggest. In the long summer days at middle latitudes, birds can incubate more constantly yet spend as many hours foraging as do tropical birds on shorter days. The northern birds' nocturnal sessions are correspondingly shorter. Moreover, most of my records of incubation constancy were made during the birds' main nesting period early in the rainy season, when afternoon downpours were frequent. While rain falls hard, birds take longer sessions on their eggs than they do in fair weather, foraging chiefly between showers. The major part of their food is gathered in the forenoon, which on many days is the only time to record incubation behavior that is not constrained by rain.

The behavior of a female Piratic Flycatcher that I watched while she incubated in a nest wrested from a pair of Gray-capped Flycatchers was instructive. During my morning vigil of six and a quarter hours, she sat with a constancy of 72.5 percent, which was high for a bird of her weight. When I resumed my watch at 2:35 P.M. she was in the nest, where she remained continuously until nightfall, without ever going off to catch a bit of supper. Yet the weather on this particular afternoon was rather better than usual in June; the sky was overcast and rain fell slowly much of the time, but with rainless intervals during which the flycatcher might have foraged without detriment to her eggs, which in any case were protected by a thick roof. Although it is not usual for a bird that takes frequent recesses in the morning to sit without interruption through an afternoon when rain is not continuously hard, the behavior of this individual was almost duplicated by that of another Piratic Flycatcher I watched in a different locality nineteen years later. During eight hours up to 3:17 P.M., she sat with a constancy of 70 percent. After this, she stayed in her nest until nightfall around 6:00 P.M., with only one absence of three minutes, hardly long enough for productive foraging. The drizzles and intermittent short showers of the late afternoon were interrupted by intervals in which the flycatcher might have foraged without a drenching.

Flycatchers, or at least some of them, take longer to break out of the eggshell than do oscine birds such as tanagers—which may be one of the factors making their incubation periods longer. The first minute fracture in the shell of a Gray-capped Flycatcher's egg, made by the bill of the chick within it, may be detected as a slight roughness of the sur-

Painted Tody-Flycatcher,
Todirostrum chrysocrotaphum.
Sexes alike. Northern South
America.

face on the second day preceding the chick's escape from it,
not on the day before hatching, as in many small songbirds.
From thirty-five to at least forty-one hours may elapse between the first indication that the chick is trying to break out
of the egg and its final release. In this interval, the chick may
be heard peeping.

As in other passerine birds, flycatchers' eggs hatch
chiefly during the night and morning. Of twenty-six eggs of
Gray-capped and Vermilion-crowned flycatchers, fourteen
hatched in the forenoon and eleven at night but only one in
the afternoon. Apparently, the imprisoned chick works to
sever its shell toward the end of the night and in the morning but rests from this strenuous labor in the afternoon and
probably also the early part of the night. A Bright-rumped
Attila's egg took thirty-eight hours to hatch. While their
eggs hatch beneath them, parents sit restlessly, often rising
up to look or poke down into the nest.

The incubation periods of flycatchers, counting from
the laying of the last egg to the hatching of this egg, range
from thirteen to twenty-three days (table 2). Only in pensile
nests, many of which are attended by females without
mates, is the period known to exceed twenty days. The small

TABLE 2   *Nest Type, Clutch Size, Incubation, and Nestling Periods of Flycatchers*

Species[a]	Nest type[b]	Clutch size[c]	Incubation period, days[c]	Nestling period, days[c]	Locality
Mistletoe Flycatcher	R	2	16–17	18–20	Costa Rica
Southern Beardless Flycatcher	R	2(3)	14–16	ca. 17	Ecuador
Mouse-colored Flycatcher	C	2	16–17	14–15	Ecuador
Mouse-colored Flycatcher	C	2	14	17	Suriname
Yellow-bellied Elaenia	C	2	15–17	17–18(21)	Costa Rica
Lesser Elaenia	C	2(1)	14–15(13)	15–16	Costa Rica
Mountain Elaenia	C	2(1)	15	14–17	Costa Rica
Torrent Flycatcher	C	2	17–18	17	Costa Rica
Ochre-bellied Flycatcher	P	2–3	19–21	18–19	Costa Rica
Tawny-crowned Pygmy-Flycatcher	C	2	14–15	11–12	Ecuador
Slate-headed Tody-Flycatcher	P	2	18–19	19–21	Costa Rica
Common Tody-Flycatcher	P	2–3	17–18	18	Surimame and Costa Rica
Spotted Tody-Flycatcher	P	2	17	17–18	Suriname
Eye-ringed Flatbill	P	2	16+	28[d]	Costa Rica
Yellow-olive Flycatcher	P	2–3	17–18	21–23(24)	Costa Rica
Golden-crowned Spadebill	C	2	18–19	14–16	Costa Rica
Royal Flycatcher	P	2	22–23	21–23	Costa Rica
Ruddy-tailed Flycatcher	P	2	22+[d]	—	Panama
Sulphur-rumped Flycatcher	P	2	22	23–24	Costa Rica
Black-tailed Flycatcher	P	2	21–22	18?[d]	Panama and Costa Rica
Bran-colored Flycatcher	C	2	17	15–17	Costa Rica
Tropical Pewee	C	3	15–16[d]	—	Costa Rica
Acadian Flycatcher	C	3(4)	14(13, 15)	13–15	USA
Alder Flycatcher	C	3–4	14–15(13)	13–15	USA
Willow Flycatcher	C	3–4	12–13?	13–14	USA
Least Flycatcher	C	4(3)	14–16(13)	14–15	USA
Hammond's Flycatcher	C	3(2, 4)	15	17–18	USA
Gray Flycatcher	C	3–4	14	16	USA
Western Flycatcher	C	3–4(5)	14–15	14–17	USA
Black Phoebe	C	4–5	16–18	21	USA
Eastern Phoebe	C	4–5(3–7)	15–16(17)	17(15–20)	USA
Vermilion Flycatcher	C	2–3	13–14	13–14(15)	Ecuador
Short-tailed Field-Flycatcher	C,R	3–5	14	12–14	Ecuador
Bright-rumped Attila	H	3–4	18	18?	Costa Rica
Great Crested Flycatcher	H	5–6(4–8)	14(13,15)	14–15	USA
Boat-billed Flycatcher	C	2–3	17–18	24	Costa Rica
Rusty-margined Flycatcher	R	2–3	16	19	Suriname
Vermilion-crowned Flycatcher	R	3–4(2)	15–16	20–22(23)	Central America
Gray-capped Flycatcher	R	2–3(4)	16–17(18)	19–21(22)	Costa Rica
Streaked Flycatcher	H	2–3	16–17	21	Panama and Costa Rica
Sulphur-bellied Flycatcher	H	3–4	16	18	Costa Rica and USA
Piratic Flycatcher	R	2–3	16	18–20	Costa Rica
Tropical Kingbird	C	2–3(4)	15–16	18–19	Central America
Snowy-throated Kingbird	C	3–4(2)	15–16	17–19	Ecuador
Eastern Kingbird	C	3–4(2)	15–16	14–17	USA
Scissor-tailed Flycatcher	C	3–5(6)	14	15?	USA

[a]Species are listed in the sequence of the sixth checklist of the American Ornithologists' Union.
[b]C=open cup amid vegetation; H=nest in hole or cranny; P=pensile nest; R=roofed nest with side entrance.
[c]Less frequent values in parentheses.
[d]Only one determination.

Short-tailed Field-Flycatcher, *Muscigralla brevicauda*. Sexes similar. Southwestern Ecuador to northern Chile.

birds that build these nests take longer to hatch their eggs than do many larger flycatchers with open nests. Likewise in the tanager family, incubation periods tend to vary inversely with the size of the birds and their eggs, those of the tiny euphonias exceeding by 25 percent or more those of the larger species. As a rule, the safer the nest, the longer birds can afford to take hatching their eggs and rearing their nestlings. Possibly pensile nests are somewhat more immune from predation than are open cups; but it is difficult to find and to learn the outcome of enough nests to prove this; certainly many of them are prematurely lost. None of the temperate zone flycatchers for which I have found reliably determined incubation periods take as long to hatch eggs as do many tropical species—which is in accord with the accelerated nesting schedules of birds of higher latitudes.

9

The Young and Their Care

A newly hatched flycatcher is a blind, helpless creature. Most hatchlings that I have seen have pink or dark skin with sparse or moderately dense natal down; but those of the Royal, Sulphur-rumped, Ruddy-tailed, and Yellow-olive flycatchers, all in hanging nests, and Golden-crowned Spadebills in open cups, are quite naked. Soon after they squirm free of the divided shell, hatchlings lift up their gaping mouths for food, revealing a yellow or orange-yellow interior. The empty shells are carried away by the parents, most often by the female.

With the exception of the male Royal Flycatcher, which guards but does not feed the nestlings, mated males nearly always bring food to their offspring and help to remove their white fecal sacs. How soon after hatching the males begin to feed depends upon how closely they have been associated with the nest while the female incubates, and probably also on their individual alertness, for, as far as I can learn, female birds have no special call to inform their partners that they have become fathers with work to do. If the male parent has made periodic visits of inspection to the nest while it contained eggs, guarded it during his mate's absences, or fed her, he is likely to start feeding the nestlings rather promptly; if he has remained aloof, days may pass before he begins. Usually he sees the nestlings before he brings food to them; but sometimes their voices appear to alert him to their presence, or seeing their mother bring food to the nest prompts him to do the same.

A male Yellow-bellied Elaenia that had helped to build the nest and had guarded it during his partner's recesses from incubation did not delay long in finding and feeding his newly hatched young. At this nest the female first brought food forty-six minutes after she left the nest containing a newly hatched nestling in the dawn. When the father first saw the hatchling, he stood for about twenty seconds looking intently down at it, then flew off carrying a white feather that stood up prominently above the nest's rim. Soon he returned with a tiny insect for the hatchling, forty-nine minutes after his mate's first morning departure from the nest, and only three minutes after she brought the hatchling its very first meal.

The events leading up to a male Gray-capped Flycatcher's discovery of his nestlings were fascinating to watch. During sixteen monotonous days of incubation, his interest in the nest had waned; but he rested much of the time in a nearby guava tree, preening, catching passing insects, and shouting to his heart's content, in continual vocal contact with his equally voluble consort. Long before sunrise on hatching day, she started to bring her nestlings insects so small that I could not see them in her bill or mouth. She came and went more frequently than on preceding mornings while she had been incubating. Her nest song, so often repeated on those mornings, was silenced by her increased activity; I did not hear it until after she had been feeding the nestlings for more than three hours. As on other mornings, she called loudly and sharply to her mate, both while in the nest and after leaving it. To me, these outcries were no different from those I had so frequently heard on earlier days. If they conveyed any new information to her partner, he did not reveal this by his behavior.

At 7:52 A.M. the female Gray-cap first brought an insect so large that I saw it projecting from her bill. Soon after this, I heard from her an utterance new to me. Standing in the doorway, back toward me, apparently coaxing a nestling to swallow an item too big for it, she voiced what sounded like a subdued version of the male's dawn song—an eager, hurried, possibly slightly vexed expostulation to nestlings slow to take their food. During the remainder of the morning, this utterance was often repeated; but apparently it was not intended to call the father's attention to his parental obligation, for I heard it at mealtime long after he had become aware of it—up at least to the nestlings' sixth day.

All these changes—increased activity at the nest, food visibly carried in the mother's bill, her attitude while delivering it, and, apparently most of all, the new sounds—aroused the male's curiosity about what was happening at the nest. One minute after I first heard the feeding song, and twelve minutes after I first clearly saw food in the female's bill, he flew past the doorway, too rapidly to see what was within. His next show of interest came when his mate, arriving with a fairly big insect that I could plainly distinguish without my binocular, alighted in front of his habitual perch, where his keen eyes could hardly fail to detect the prey. Now he flew down to hover above her while she fed the nestlings, her back in the doorway hiding them from him. After her departure he poised twice beside the nest and once above it, but not directly in front, where he could clearly see inside— the greatest display of interest that I had witnessed during fifteen hours of watching. He was obviously trying to learn what was happening at the nest but, perhaps a young bird with no previous experience of parenthood, he pursued his investigations so ineptly that only after hovering many times in front of the entrance, while his consort was delivering food or immediately afterward, did he achieve success.

Finally, seventy minutes after the arousal of the male Gray-cap's curiosity, he alighted on the doorsill and delayed there long enough to scrutinize the babies within. His response to the sight was unhesitating; he knew precisely what to do, and he did it in the most approved manner. Going to a nearby tree, he flew out to seize the first suitable insect that winged by, mashed it in his bill as is proper for hatchlings, and within two minutes was back to feed one of them, whispering a snatch of the nest song while he did so. This occurred four hours and thirty-two minutes after his mate's first morning departure revealed the hatchlings. She had brought their first meal only twenty-four minutes after her day began.

I have dwelt in so much detail upon the male Gray-cap's slowly achieved discovery of his offspring because, more than any other bird that I have watched on hatching day, he appeared to try hard to solve a mystery; most of the others have, early or late, by chance or design, become aware of their nestlings at a stroke, as the Yellow-bellied Elaenia did. At another Gray-cap's nest, where the mother started to feed the nestlings only twenty-six minutes after her earliest opportunity to do so, the male began after more than six hours but less than thirty-two hours. A female Vermilion-crowned Flycatcher brought food thirty-two minutes after her first nestling hatched; its father delayed six hours and thirty-four minutes. Another female Vermilion-crown brought food almost immediately after her young hatched; her unobservant partner, not until between six and ten days later. A female Bran-colored Flycatcher started to feed twenty-seven minutes or less after her nestling emerged from the shell; her mate, between four and twenty-nine hours thereafter.

In less than twenty-four hours (including the night) after three Dusky-capped Flycatchers hatched in a hollow bamboo in a garden trellis, the father hesitantly entered the deep cavity with a moth in his bill. Since he had not been in the habit of entering to inspect the nest, I surmised that he was prompted to bring food by seeing his partner do so. A female Bright-rumped Attila brought the first meal about one and a half hours after her first nestling hatched; her mate, between five and nine hours after that hatching. Between five and seven days after nestlings of the Yellow-olive Flycatcher hatched, their father started to feed them. He was not seen to enter the retort-shaped nest during incubation, and without going in he could not see what it held. In this case, too, he probably brought food because he saw his mate, with whom he was closely associated, arriving with food in her bill.

Flycatchers that include little or no fruit in their diet of insects and spiders give little or none to their nestlings. With admirable adroitness, flycatchers add insects to a bill already laden with them before they approach the nest with their harvest. They pound large or hard insects against a perch or mash these in their bills before delivery, but sometimes they give a nestling dragonflies with widespread filmy wings still attached, which the young gulp down with difficulty. Flycatchers that eat many berries and arillate seeds bring them increasingly to their growing nestlings. Among the more frugivorous species is the Ochre-bellied Flycatcher, which regurgitates food to its nestlings. All other flycatchers, as far as I know, feed directly from the bill. Nestlings slow to swallow their meals may be coaxed with low notes not unlike the nest song, but softer. When a male arriving with food finds his partner brooding the nestlings, he passes it to her and she rises up to give it to them, or he may wait until she leaves them uncovered and he can feed them directly.

After digesting off the pulp of the fruits they receive, nestlings regurgitate the seeds. Mistletoe Flycatchers, which eat many drupes of the parasitic plant for which they are named, often arrive at the nest with a whole row of fruits lined up in their short black bills. The nestlings continually eject through their mouths seeds still covered by the inner mucilaginous layer that attaches them to branches of living trees, where they germinate. Sometimes a young bird promptly reswallows the seeds that it has regurgitated; but many adhere to its lower mandible or chin, where two or three may hang together until a parent removes them after delivering food. Other seeds fall directly through the doorway in the nest's side, lodge at the entrance, or are wiped off there by the nestlings. Parent Mistletoe Flycatchers swallow seeds ejected by their young; possibly, with more efficient digestion, adults derive additional nourishment from them. Bright-rumped Attilas, which from the day they hatch receive many lizards, when older sit with two or three inches of lizard's tail projecting from their mouths while they digest the small reptile's anterior parts. Another tail may hang in front of their brooding mother.

Hourly rates of feeding nestlings by twenty-one species of flycatchers that I have watched in Central America are given in table 3. In successive hours the number of meals may fluctuate greatly because an interval of rapid feeding of hungry nestlings is often followed by one of reduced frequency after they are satiated. On longer watches, these fluctuations level off. Flycatchers that bring their nestlings many small insects, perhaps interspersed with a few berries, achieve the highest rates, up to thirteen to eighteen times per hour for each nestling by Torrent, Sulphur-rumped, Yellow-olive, Gray-capped, and Vermilion-crowned flycatchers and Yellow-bellied Elaenias with older nestlings. Solitary females may feed about as rapidly as pairs because the rate of provisioning is determined primarily by the nestlings' needs rather than by the parents' ability to bring meals.

Boat-billed Flycatchers compensate for infrequent feeding by bringing large items, often cicadas so large that the nestling can gulp them down only with an effort. Meals brought by Bright-rumped Attilas are few but substantial. Like other birds that feed their young by regurgitation, Ochre-bellied Flycatchers deliver meals that are widely spaced but apparently ample. In contrast to the rapid deliv-

TABLE 3　　*Rates of Feeding Nestling Flycatchers*

Species	Number of nestlings	Age in days	Hours watched	Male	Female	Total[a]	Meals per nestling per hour
				Number of feedings			
Yellow-bellied Elaenia	1	6–7	8	33	28	77	9.6
Yellow-bellied Elaenia	1	16	4	+[b]	+	59	14.8
Yellow-bellied Elaenia	2	16, 17	4	+	+	107	13.4
Torrent Flycatcher	2	ca. 15	3	+	+	110	18.3
Ochre-bellied Flycatcher	3	ca. 5	4	0	7	7	0.6
Ochre-bellied Flycatcher	3	ca. 15	4	0	8	8	0.7
Scaly-crested Pygmy-Flycatcher	2	ca. 15	4	0	57	57	7.1
Northern Bentbill	2	ca. 10	8	0	87	87	5.4
Eye-ringed Flatbill	1	ca. 7	7	0	10	10	1.4
Yellow-olive Flycatcher	1	19–23	4 ½	+	+	65	14.4
Golden-crowned Spadebill	2	7, 8	3 ½	+	+	46	6.6
Royal Flycatcher	2	5	4	0	8	8	1.0
Royal Flycatcher	2	19	2	0	14	14	3.5
Ruddy-tailed Flycatcher	2	3	4	0	10	10	1.25
Sulphur-rumped Flycatcher	2	6,7	4	0	54	54	6.8
Sulphur-rumped Flycatcher	2	ca. 20	2	0	66	66	16.5
Bright-rumped Attila	3	3	5	+	+	5	0.3
Bright-rumped Attila	4	4	5	+	+	13	0.7
Bright-rumped Attila	4	ca. 7	9 ½	+	+	27	0.7
Bright-rumped Attila	4	ca. 17	5	+	+	16	0.8
Dusky-capped Flycatcher	3	5	4 ⅓	7	7	16	1.1
Boat-billed Flycatcher	1	ca. 20	5 ⅔	6	5	25	4.4
Vermilion-crowned Flycatcher	1	10	3	17	22	39	13.0
Vermilion-crowned Flycatcher	1	19–21	8	+	+	104	13.0
Gray-capped Flycatcher	2	17	6	+	+	158	13.2
Tropical Kingbird	2	9	4	16	24	40	5.0
Tropical Kingbird	2	17	4	19	20	42	5.3

[a]The total number of feedings may be greater than the sum of feedings credited to each parent because sometimes the feeder's sex was not recognized.
[b]Plus signs indicate that both sexes fed the nestlings but were too similar in appearance to be distinguished.

ery by birds feeding nestlings directly from the bill, Ochre-bellied Flycatchers take up to two minutes to deliver a meal, the rather violent movements of their bodies clearly indicating that they regurgitate, as in hummingbirds. The most rapid provisioning that I have witnessed was by a pair of Gray-capped Flycatchers, who in one hour brought food sixty-five times to three eleven-day-old nestlings.

While I studied a group of birds whose nests had rarely, if ever, been watched, I deemed it more valuable to learn the main features of their life histories than to accumulate great masses of data. Others, working with birds that were already much better known, have concentrated on one or a few species, often with teams of observers or automatic recording devices, making far more detailed studies of flycatchers' rates of feeding their young. Unfortunately, the diverse ways that their data are presented, and often the sheer mass of data, make it difficult to fit the information into a single table. Among the more prolonged of these studies is that by S. Charles Kendeigh, who learned that a pair of Eastern Phoebes, feeding young twelve days old, made 845 trips to their nest in a single day. After this maximum, the rate declined slightly until the young phoebes' departure at the age of seventeen days.

In an exceptionally detailed study of the Western Flycatcher, John Davis, George Fisler, and Betty Davis learned that the sexes brought almost equal numbers of meals to their nestlings, the male parents contributing a slightly larger share at two nests, the females at two others. The greatest differences were 57.4 percent of feedings by the male to 42.6 by the female at one nest, and 44.6 and 55.4 percent by the male and female, respectively, at another. Likewise, Elmer Morehouse and Richard Brewer, in a prolonged study of the Eastern Kingbird, found that the two parents contributed about equally to their nestlings' fare, the father bringing more meals on some days, the mother on other days. In contrast to this, during eighteen hours of watching by Frank W. Fitch at Scissor-tailed Flycatchers' nests with three or four young, the females brought food ninety-one times and the males only thirty times. A much greater disparity in the contributions of the two parents was disclosed by David E. Davis, who in 16.5 hours of watching at four Hammond's Flycatcher nests saw females bring food 122 times, the males only seventeen times.

Although polygamy is rare (or at least rarely discovered) in the many species of flycatchers that normally nest as cooperating pairs, a few cases have been reported in migratory species in temperate North America, where flycatchers have been most intensively studied. A small minority of Willow Flycatchers bred with two consorts in localities as widely separated as Oregon, Colorado, and Ontario. Other species of *Empidonax* in which a bigamous male or two have been detected are the Least and Acadian flycatchers. Single

instances of bigamy have been noticed in the Eastern
Phoebe, Western Wood-Pewee, and Eastern Wood-Pewee.
Except possibly in the last-mentioned, all these males with
two wives have fed the young of both, although usually
much less than the mothers brought to their broods. Most
often the two females mated to a single male nested in well-
separated parts of his territory, but the two consorts of one
Acadian Flycatcher had nests in territories separated by
that of a different male.

In normally monogamous birds, polygamy may arise
because males are fewer than females, or it may result from
territorial fidelity. A female, returning to the spot where she
nested with success in the preceding year, may find it occupied by an already mated male other than her last-year's
partner, who has failed to return. Her strong attachment to
the familiar site may lead her to settle there, at the price of re-
ceiving little help in rearing her young from a male that has
other obligations. Usually the two consorts of these bigamous flycatchers lived amicably together. One female Willow Flycatcher was seen to fly from her eggs to feed the other
female's fledgling, begging in a nearby shrub. In California,
Michael Stafford watched two or more unrelated Willow
Flycatchers and one Dusky Flycatcher help the mother of
four fledgling Willow Flycatchers to feed them.

In cooperative breeding associations, a mated pair is as-
sisted in defending territory and rearing a brood by helpers
that are usually their older offspring but sometimes less
closely related. Such groups have rarely been found among
flycatchers, perhaps only because such a small proportion
of species have been carefully watched. The most thorough
study of a cooperatively breeding flycatcher was made by
Betsy Thomas on the Venezuelan *llanos*, wide, grassy plains
with scattered palms and other trees, swept by drying winds
for six hot, rainless months and flooded by tropical downpours during the rest of the year. Here White-bearded Flycatchers live in pairs or trios, small birds with black caps
separated by white eyebrows from black cheeks, olive-

brown upper parts, white throats, and bright yellow breasts and bellies. Their neat, shallow, lichen-covered cups, not unlike the nests of Eastern Wood-Pewees and Yellow-bellied Elaenias, blend with the bark of small trees where they are built in horizontal forks. From early April to mid-July, females lay sets of two white eggs, sparsely spotted and blotched with chestnut. Of three carefully watched nests, one belonged to an unaided pair, and each of the other two was attended by a trio. At each of the nests with a helper, all three adults built, fed the nestlings, and defended the territory. Two of them shared incubation and brooding rather equally; the third incubated only occasionally and apparently not very effectively. The helpers brought about one quarter of the nestlings' meals and also carried away their droppings. When, after seventeen or eighteen days of incubation and a nestling period of eighteen days, the young left the nest, the helper was not seen to feed them, although it participated in the defense of the fledglings and the territory. One young White-bearded Flycatcher continued to follow and beg from its mother until it was three months old, although at half this age it caught insects for itself.

Like other cooperative breeders, these White-bearded Flycatchers lived amicably together. They duetted while they seesawed up and down and rhythmically flapped their wings. Sometimes one joined in a duet while sitting in the nest. Juveniles preened each other. After the young fledged, all members of a group foraged and roosted close together. The male's participation in incubation and brooding differs so strikingly from the known behavior of other flycatchers that one wonders whether it is usual even in this species.

In plumage, Rusty-margined Flycatchers, widespread in tropical South America, resemble White-bearded Flycatchers and, even more closely, Vermilion-crowned Flycatchers. Their domed nests with a side entrance are easily confused with those of Vermilion-crowned and Gray-capped flycatchers. The only Rusty-margined Flycatcher nest at which helpers have been found was situated on a branch of a dead tree that had fallen into Gatún Lake in the former Panama Canal Zone, near the northern limit of the species' range. Here Robert Ricklefs watched four adults, alike in appearance, bring food to three half-grown nestlings 211 times in six hours. The attendants usually approached the nest together, and one fed the nestlings while one or more of the others watched. While delivering food in a roofed nest, the feeder's back is exposed, but with its head inside it cannot see an approaching predator. Cooperative feeding by these Rusty-margined Flycatchers might reduce the risk of predation; while one member of the group feeds the nestlings, others serve as sentinels to warn of danger.

Among flycatchers, as among other birds, males that do not incubate scarcely ever brood the nestlings, for brooding is essentially the same activity as incubating. Especially when the father begins promptly to bring food, the mother may brood her hatchlings almost as constantly as she incubated the eggs from which they hatched. As the nestlings' demand for food increases, she devotes more time to foraging and less to covering them. After their feathers expand and they develop some capacity to regulate their own temperature, she may no longer brood them in fair weather, but hard rain sends her to cover them days after daytime brooding has ceased. When sunshine falls strongly on the nest, she may stand above the nestlings, shading them, instead of sitting over them.

For some days after daytime brooding in fair weather has been discontinued, mothers cover well-feathered nestlings through the night. Often the young sleep alone during their last nights in the nest, but some careful parents warm them on the final night, as Lesser Elaenias sometimes do. Yellow-olive Flycatchers, which use their pendent nests as dormitories, accompany their nestlings by night as long as they remain in the nest, then continue to sleep there alone, leaving their fledglings to roost amid foliage. In their roofed nests, Vermilion-crowned mothers brood their nestlings by night until they have grown so big, and occupy so much of the available space, that she sits far forward, in what appears to be an uncomfortable posture, with her head exposed in the doorway instead of turned back and snuggled among her plumage. Even on a sunny morning, I have watched a Boat-billed mother trying to brood a nestling so big and well feathered that it certainly did not need to be covered. Appearing not to relish being sat upon, it was so restless that at intervals the parent would leave the nest and perch nearby, guarding. Evidently, the young Boat-bill was brooded as an alternative mode of guarding it, not because it needed to be covered.

While still in the nest, some young flycatchers, including the Gray-cap, Vermilion-crown, and Mistletoe, repeat the calls and even songs of the parents but in weaker voices; they may become quite noisy. Such precocious utterance of the adults' vocalizations, widespread in this and other nonoscine families, is rare or absent in true songbirds, which do not practice the more complex songs of adults until some weeks after they have left the nest. Songbirds learn their songs from older individuals; but the songs of Willow Flycatchers, Alder Flycatchers, Eastern Phoebes, and probably other flycatchers are innate.

If undisturbed, flycatchers remain in their nests until they are well feathered and can fly. A day or two before their departure, those in open cups may venture over surrounding branches for a few inches, then promptly return. They exercise their wings by beating them rapidly. Nearly always, unless danger threatens, they fly from their nests spontaneously, with no parental urging. All may exit within a few minutes, or nestmates may leave on successive days. If a parent is in view as a fledgling wings unsteadily from the nest, the adult may escort it in "shielding flight," close above or slightly behind it, in a position to deflect an attacking raptor from the weakly flying youngster to itself, an adult better able to escape. When a Tropical Kingbird ended its first flight on an exposed twig, the parent who had been shielding it promptly knocked it down into concealing foliage.

Flycatchers abandon their nests at ages ranging from twelve to twenty-four days, rarely longer (table 2). Those reared in open nests tend to leave at an earlier age than do those in roofed nests. Among the longest nestling periods are those of species that build hanging nests and also have exceptionally long incubation periods. Slow embryonic development is followed by slow growth and maturation; but the safety of nests dangling beyond reach of most flightless predators makes it advantageous for young birds to remain inside until they are strong. When the parents are alike, fledglings leave their nests closely resembling the adults but with shorter tails. Indeed, their fresh new plumage may be brighter than the worn and faded feathers of parents that for weeks have toiled to rear them. Although the sexes of adult Royal Flycatchers differ only in the more intense color of the male's crest, in juvenal plumage they are more strongly barred. When the sexes differ greatly in plumage, as in the Vermilion Flycatcher and some South American species, fledglings tend to resemble their mothers more than their fathers, as is usual in birds. When adults have very long streamers, as in Scissor-tailed and Fork-tailed flycatchers, juveniles may be recognized by their shorter tails and duller plumage.

After quitting the nest, fledglings follow their parents, often beyond the parental territory, and in their stationary intervals perch in contact while waiting to be fed. At night siblings roost pressed together, often with a parent sleeping nearby or, as sometimes happens in Common Tody-Flycatchers, in the same compact row.

Although it is difficult to keep mobile families with fledged young in plain view long enough to make significant records of the rate of feeding, Morehouse and Brewer succeeded in following a family of Eastern Kingbirds for forty days. They learned that the rate of feeding reached a peak in the second week after the young left the nest, when one brood was fed up to 10.3 meals per young per hour, almost twice the maximum they received while in the nest.

Likewise, the Western Flycatchers studied by John Davis and his team were fed more rapidly a few days after they left the nest than they had been before they flew. They did not flutter their wings while being fed, as young songbirds often do, but simply turned toward the parent, who either alighted beside them or poised in front to deliver a meal. Likewise, a Yellow-bellied Elaenia was fed slightly more rapidly on the day after it left the nest than during nineteen hours when it was observed as a nestling.

Seven or eight days after they abandon their nests, young as different in size as the Western and Acadian flycatchers and the Eastern Kingbird can capture insects in flight, and thereafter their skill in flycatching increases rapidly. As they become better able to nourish themselves, parental ministrations decrease. Acadian Flycatchers were fed by their parents for at least two weeks after their first flight; Western and Least flycatchers for two or three weeks; Eastern Phoebes for fourteen to twenty-two days. The last feeding of a brood of Eastern Kingbirds was seen on their thirty-fifth day out of the nest. As in other tropical birds, tropical flycatchers feed their young to a more advanced age. In François Haverschmidt's garden near Paramaribo, Suriname, two Common Tody-Flycatchers left their swinging nest when eighteen days old. Twenty-seven days after they flew, one was first seen to catch an insect for itself. Then followed an interval when self-feeding was supplemented by parental contributions. These young tody-flycatchers were last seen to receive food from a parent forty-three days after they left their nest, when they were sixty-one days of age.

Families of nonmigratory tropical flycatchers may remain intact long after parental feeding ceases. Young Boat-billed Flycatchers stay with their parents for up to eight months after they fledge, or until preparations for nesting begin in the following year. Groups of three, four, or five of these yellow-breasted birds straggle through the treetops one after another instead of flying in a compact flock, covering long distances in the course of a day, and repeating their drawled *churr* while they rest between flights.

The number of broods, or of nesting attempts, is constrained largely by the length of the favorable season. In Canada and the northern United States, where parent flycatchers have only a few months to rear their young and leave time for all members of the family to prepare themselves for the long, energy-demanding southward migration, second broods appear to be rare, and replacement nests may not be undertaken if the first brood is lost at an advanced stage. In the southern United States, with a longer nesting season, more broods can be reared. In temperate North America as a whole, second broods are occasional in a number of small *Empidonax* flycatchers, including the Least, Western, Dusky, Gray, and Acadian. In Michigan, a female Acadian Flycatcher started to build her second nest while feeding nestlings of her first brood. Renesting after nest failure is more frequent.

All three species of early-nesting phoebes, the Eastern, Black, and Say's, often rear two broods and sometimes three. The day the last nestling of an Eastern Phoebe's first brood left the nest, the first egg of a second clutch of five was present in it. Eastern Kingbirds and Scissor-tailed Flycatchers appear seldom or never to attempt second broods, but Cassin's Kingbird does in the South. After losing two nests, a pair of Vermilion Flycatchers in southern California reared a brood, then laid three eggs for another, the outcome of which is not recorded. This was their fourth nesting attempt between February and June.

Here in the valley of El General, at nine degrees north and two thousand to three thousand feet above sea level, most birds breed early in the rainy season from March or April to June or July. Yellow-bellied and Lesser elaenias sometimes resume laying twelve or thirteen days after their first broods leave the nest, and Mistletoe Flycatchers sometimes rear second broods. The few times that I have found any other flycatchers in this area trying to rear second broods,

Pied Water-Flycatcher, *Fluvicola pica.* Sexes similar. Eastern Panama and northern South America to central Argentina, Trinidad.

they failed. At lower altitudes, and nearer the equator, breeding seasons are longer and more broods are started. In the lowlands of Suriname, at five degrees north, several flycatcher species are engaged in nesting through most or all of the year. Among them Haverschmidt named Pied Water-Flycatchers, Great Kiskadees, Rusty-margined Flycatchers, and Spotted Tody-Flycatchers. It is doubtful that one pair could engage continuously in reproduction for so many months; more probably, different pairs nest at different times in a year almost continuously favorable. A pair of Rusty-margined Flycatchers reared three broods, each containing two fledglings, in five months. While still feeding nestlings of the first brood, they started to build a nest for the second. After rearing the third brood, they rested for six months before nesting again. From April to January, a pair of Spotted Tody-Flycatchers built ten nests in Haverschmidt's garden and laid an egg or two in each of them, but they lost all to nest robbers and failed to rear a single young. In the rainier parts of the tropics, where predators abound and so many nests fail, flycatchers try again and again to rear a brood, but probably few succeed in producing more than one in a season.

10

Enemies, Defense, Nesting Success, and Longevity

Birds as vigilant as kingbirds and other large flycatchers are probably rarely surprised by hawks and other diurnal predators. However, even the biggest and boldest of them are occasionally ambushed. William Beebe told how a Great Kiskadee fishing from a sandbar in Mexico was seized and eaten by a raccoon that stealthily approached to within a short distance, then rushed upon the preoccupied bird, whose wet plumage impeded swift flight. The poor bird screamed piercingly before it expired. At night flycatchers are probably no safer from owls, opossums, weasels, snakes, and other nocturnal prowlers than are other diurnal birds of no great size.

More vulnerable than adult birds, eggs and nestlings are taken by a host of hungry creatures, flying and flightless. Against such large, conspicuous diurnal nest robbers as hawks, kites, caracaras, crows and jays, and toucans in wooded regions of tropical America, kingbirds and other large flycatchers of open and semi-open country have adopted a foresighted policy of aggressive defense. They do not wait for one of these pillagers to approach a zealously guarded nest. Spying it from afar and shouting their battle cry, the flycatchers sally forth to greet the predator and drive it to a still safer distance. Such is the energy of their pursuits that much bigger birds flee, often not swiftly enough to avoid being pounced upon and losing feathers from their backs.

With toucans, this strategy is especially appropriate, for in the air these heavy fliers cannot defend themselves, but after they alight beside a nest and can turn their heads around, not the boldest flycatcher dares to confront their great, brightly colored, menacing bills. Boat-billed Fly-catchers are ever ready to pursue big Chestnut-mandibled Toucans, which are among the most frequent despoilers of Boat-bill nests. In the Costa Rican highlands, Dark Pewees become intensely excited whenever an Emerald Toucanet alights in view of their nests but hesitate to come to grips with the much bigger birds until they start to fly away, when the peewees can safely dart close enough to buffet them. Females leave their nests to join their mates in chas-ing toucans. These woodland predators deserve all the pun-ishment that flycatchers give them. One day I surprised a Chestnut-mandibled Toucan in the act of gulping down a well-feathered, two-week-old Vermilion-crowned Flycatcher nestling. To reach its victim, the huge-beaked bird had torn off the whole roof of the flycatcher's nest.

As strong as the antipathy of Boat-bills to the larger toucans and of Dark Pewees to the smaller toucanets is that of Tropical Kingbirds to Swallow-tailed Kites. In Costa Rica, these graceful birds subsist chiefly upon insects that they catch with their feet while they soar high in the air on as-cending currents. They also tear nestlings from exposed nests, such as Tropical Kingbirds often build, that kites can reach while hovering. Any large bird passing within a few hundred feet of a kingbird's nest, especially if unfamiliar, is likely to evoke spirited attack. Among winged things assailed by Eastern Kingbirds were an egret, a tern, and a slow airplane flying low.

One of the first flycatchers' nests that I studied was that of a pair of Vermilion-crowns in a citrus tree in a Panamanian garden. As I approached on a dark and rainy noon, a Black Hawk-Eagle rose from the tree with two feathered nestlings clutched in its talons, along with the straws from their roof, all in one cruel embrace. The raptor flew with its victims to a neighboring fig tree where, surprisingly, it perched for an hour, making no move to devour the nestlings that hung limply from its grasp until carrion-eating insects buzzed around them. Resting there, the raptor was surrounded by the bereaved parents and their neighbors, Gray-capped Flycatchers and Garden Thrushes, all darting so close to its head that it ducked, the Vermilion-crowns complaining in querulous tones, the Gray-caps protesting with a staccato *wic wic wic*, the thrushes uttering melancholy cries very different from their lilting songs. After a while the flycatchers drifted away, leaving three or four thrushes to continue their spirited offensive with darts that often grazed the hawk's head.

One would expect kingbirds, Boat-bills, and other flycatchers that rout hawks, crows, toucans, and other birds much bigger than themselves to be invincible against innocuous birds of about their own size or smaller, but occasionally they are resisted or even defeated. Their assaults on flying predators might be compared to a military airplane's attack on a much larger passenger or cargo plane—the maneuverability of the flycatchers gives them a great advantage. In confrontations with small perching birds, kingbirds and other flycatchers, for all their reputed pugnacity, demonstrate no superiority and frequently lose. Eastern Kingbirds have been repulsed by a Yellow Warbler, a young Yellow-bellied Sapsucker, a Baltimore Oriole, and an American Robin. A building pair lost their nest site to an aggressive male House Sparrow. A pair of Great Crested Flycatchers failed to save the gourd in which they nested from Purple Martins who wanted it for their eggs and young. Despite the parent flycatchers' attacks, a martin pulled a tender nestling from the gourd and dropped it to the ground.

The male of a pair of Sulphur-bellied Flycatchers nesting in a cavity in a tall dead trunk frequently pursued harmless Resplendent Quetzals feeding nestlings in a neighboring hole, worrying them without hurting them. He drove away a Ruddy Pigeon innocently eating berries in the top of a tall tree. When he darted toward a pair of Spangled-cheeked Tanagers that alighted in a tree in front of his nest, one of these much smaller birds held its ground, screaming shrilly, while the assailant hovered momentarily above it. Finding that the tanager refused to be driven away, the flycatcher desisted from menacing it; and for several minutes the pair of tanagers and the pair of flycatchers rested peacefully close together in front of the nest. Similarly, a male Tropical Kingbird that tried to chase a Golden-fronted Woodpecker from near his mate's nest desisted when the woodpecker clung beneath a branch and looked defiantly up at the flycatcher, who fearlessly chased hawks. Flycatchers rarely press attacks upon small, harmless birds that resist them, and they often permit such birds to rest close to their nests without protest. The Tropical Kingbird resisted by the woodpecker allowed doves, seedeaters, and even Groove-billed Anis to perch in his nest tree.

Although the smaller flycatchers rarely undertake long aerial pursuits of predators, they valiantly confront intruders into the precincts of their nests, often with an appearance of fury that compensates for their puny size and has given them a reputation for unrestrained belligerence. A Willow Flycatcher drove a Red-winged Blackbird from near its nest, pecking its head and neck much as larger flycatchers chastise hawks and toucans.

With swift onslaughts and loudly clacking bills, flycatchers chase squirrels and other small, nest-plundering mammals from near their nests; even humans are not immune from their assaults. By spreading its crown feathers to display the bright red or yellow patch on top of its head as it dives swiftly toward an intruder, a flycatcher may make its attack more fearsome. Dusky-capped Flycatchers, after I had inadvertently driven their well-feathered nestlings from their hole in a low stub, swooped with sharply clacking bills so close to my face that, despite my intention to remain erect and open-eyed to see how near they would come, I ducked my head to protect my vision. Both sexes of a pair of tiny Tropical Pewees dived down at my head with clacking bills and a little, trilled, protesting *cheee* as I studied their nest. By the same tactics, they drove away a pair of Garden Thrushes, a Black-striped Sparrow, a Vermilion-crowned Flycatcher, a Yellow-bellied Elaenia, and even a Great Kiskadee, all birds bigger than themselves. Bill snapping is an effective way of distancing intruders; by this means a diminutive Black-headed Tody-Flycatcher drove from its nest a relatively huge Great Kiskadee.

The domineering Great Kiskadee commands respect. W. H. Hudson recounts how a pair that nested annually in his orchard in Argentina darted down to strike his head with beak and wings whenever he ventured near the peach tree that held their large, untidy nest. When I examined a Vermilion-crowned Flycatcher's nest in a palm tree, a kiskadee with a higher nest on a nearby telephone pole joined the parent Vermilion-crowns in threatening me. Thereafter, whenever I passed near their nest wearing the hat I had worn on that occasion, one member of the kiskadee pair would dart down at me, displaying its yellow crown patch to intensify the effect. Without the hat, I could pass below the nest unthreatened, as could other people who showed no interest in the birds.

The big Boat-billed Flycatcher, which flies far to drive nest-plundering toucans still farther from its nest, is equally protective of its fledglings. After three young Boat-bills left their nest in a cecropia tree behind my house, the parents were so watchful and cautious that I could not come within a hundred yards without stirring up a storm of protests. If I suddenly emerged from the forest adjoining the pasture, an anxious Boat-bill would notice me almost immediately and fly up to scold. Then one or both of the parents would follow me about the pasture, often swooping angrily over my head and resting on low boughs of guava trees to complain with drawled churrs. All this while the objects of their concern would remain out of sight. No flycatcher has ever struck me, but both Eastern and Western kingbirds hit Michael Murphy repeatedly while he weighed their nestlings.

In the tropics and temperate zones, snakes are the most frequent despoilers of birds' nests. One morning in Honduras, I watched a seven-foot yellow-and-black mica, a serpent that preys insatiably upon eggs and nestlings, climb upward through the tangle of vines that enveloped the lower half of a tall tree trunk. A long stretch of clean bole separated the vines from the tree's spreading crown, where eighty long pouches woven by Montezuma Oropendolas hung like huge gourds. If the snake could reach them, it might lurk in the tree until it had devoured all the big birds' eggs and nestlings. Although some snakes are adept at ascending branchless trunks that appear too smooth for them, I was not sure that it could scale this one.

Fortunately for the oropendolas, the mica was not given an opportunity to test its skill as a climber. A pair of Great Kiskadees with a nest in a lower, nearly leafless tree about 30 feet away, could not tolerate so formidable an enemy so close to their young. Repeatedly they shot past the big snake's head, sometimes missing it by barely an inch, voicing their shrill *eeek* when closest. A few such spirited demonstrations halted the snake's ascent; a few more made it turn and twine slowly downward through the tangle. As long as the reptile was visible, the kiskadees continued to threaten it, just as they menaced me when I visited their nests. The oropendolas, who had so much to lose, gave little attention to the commotion below, possibly because they felt secure above the long, smooth upper half of the trunk.

As oropendolas have discovered, the success of nests depends largely upon their placement. Although Tropical Kingbirds rely upon active defense rather than concealment, Eastern Kingbirds are more prudent. A study by Michael Murphy revealed that in both New York and Kansas, their nests were most likely to escape loss when situated at about mid-height of trees and about half-way between a tree's center and the outside of its canopy; and at such sites their nests were most often placed. Screening foliage and adequate supporting branches also increased success.

Distraction displays, including injury simulation, by which birds from little wood warblers to great Ostriches lure enemies from their eggs and young, are rarely given by flycatchers. Their strategy is to repulse predators from their nests, not to entice them away with the prospect of a meal that the hungry mammals rarely catch. Among the few reported to employ this ruse is the Gray Flycatcher, who dives quickly to the ground when her nest is approached, fluttering and bouncing buoyantly away with one or both wings loosely drooping, appearing too crippled to fly. Ned Johnson saw females display in this manner on five of nine visits to their nests. He watched Dusky Flycatchers "feign injury" on only two of the eleven occasions when he put them off their nests. Other species of *Empidonax* have been carefully watched without revealing injury simulation. As Miguel Alvarez del Toro approached a nest of the rare little Belted Flycatcher in Chiapas, Mexico, the parent dropped from the nestlings she was brooding and fluttered away as though crippled. When I visited a low nest of a Lesser Elaenia, she jumped from her eggs and fluttered slowly a few inches above the ground. This not very convincing simulation of an impediment is the only one that I have seen in many hundreds of visits to nests of over forty species of flycatchers.

Instead of attempting to drive intruders from their nests or to entice them away by simulating injury, some birds simply continue to sit tight, trying to escape notice by avoidance of revealing movement. Such extraordinary "tameness" is often displayed by small *Empidonax* flycatchers. Least, Gray, and Hammond's flycatchers have permitted themselves to be lifted from their nests by people who could not otherwise learn what they covered. Even more amazingly, a Least Flycatcher clung to her nest while the small tree in which she incubated was cut down. I have not known any tropical flycatcher to exhibit such supreme steadfastness, but a Mistletoe Flycatcher remained sitting on her hatching eggs while I climbed a ladder and touched her bill. While incubating, Bright-rumped Attilas, large and very active birds, have permitted me to come within arm's length. Eastern Phoebes, which often build their nests in human habitations, become almost indifferent to the presence of people.

In my experience, flycatchers with pensile nests try neither to drive nor to entice away potential predators, but for safety they rely upon the inaccessibility of their structures. For added security, they frequently hang these close to nests occupied by stinging wasps. Of fifty-two nests of Yellow-breasted Flycatchers found in Suriname by François Haverschmidt, thirty-nine were close to vespiaries. In the same country, Gray-crowned Flycatchers, Painted Tody-Flycatchers, and Spotted Tody-Flycatchers also hang their nests near these insects. In Costa Rica, Yellow-olive Flycatchers frequently suspend their nests within a foot of a vespiary; but if none is readily available, they choose as a substitute a silken nest woven amid foliage by ants *(Camponotus senex)* that swarm over the disturber of their homes but lack a sting.

Wasps do not always provide adequate protection; in Ecuador, S. Marchant watched a snake extract two newly laid eggs from a Southern Beardless Flycatcher's enclosed nest with a side entrance, only thirty-one inches from a small vespiary. Probably the gliding snake did not shake the supporting branch enough to arouse the wasps. These insects sometimes establish themselves inside a bird's nest near one of their own, usually after its abandonment by the birds. I have repeatedly found them in the retort-shaped structures of Yellow-olive Flycatchers. Roofed nests with side entrances, such as those of the Gray-capped Flycatcher and the Great Kiskadee, are frequently situated near vespiaries; but birds with open nests do well to avoid proximity to stinging insects.

In more open regions of Mexico and Central America, but not in rain forests, grow small acacia trees or shrubs that in paired, swollen thorns harbor stinging ants of the genus *Pseudomyrmex*. The acacias nourish their tenants with tiny protein corpuscles abundantly produced at the tips of myriad tiny leaflets. In these bull's-horn acacias a diversity of birds including flycatchers, wrens, and orioles, build domed or pendent nests. The ants provide very effective garrisons, as anyone who has tried to examine the contents of nests so situated can attest. A small mammal or reptile placed in their tree is promptly routed by the stinging insects. Sometimes a bird's nest in a bull's-horn acacia is built close beside a wasps' nest, to have both flying and flightless guardians.

Dependence upon irascible insects for protection is not without a special peril. On the savannas of Nicaragua, as Thomas Belt told, a "yellow and brown flycatcher" (probably a Great Kiskadee or a species of *Myiozetetes*) builds its domed nest in a shrub armed with curved prickles, often beside the nest of a banded wasp. As he and his party rode past, a flycatcher darted out of her nest and was caught just under her bill by one of the hornlike thorns. While trying to extricate herself, she was further entangled. Aroused by her wild fluttering, the wasps attacked her. Members of Belt's party who tried to rescue the bird were also assailed by the insects, and one was severely stung. In less than a minute, the flycatcher died from the stings and was perforce left hanging in front of her nest, while her mate flew round and round screaming in terror and distress.

Among the smaller enemies of nesting flycatchers (and

many other birds) are *tórsalos,* the larvae of a dipterous fly (*Philornis*) that develop just under the nearly naked skin of nestlings, where they form conspicuous swellings, principally on the head, back, and wings. Often six or eight of these larvae infest a single host. I kept watch over one nestling Vermilion-crowned Flycatcher that supported eleven *tórsalos* and appeared to be none the worse for the infection. But the plumage of three heavily parasitized Graycap siblings developed abnormally slowly. One by one they vanished from the nest—whether removed by the parents after they succumbed or carried off by a predator, I could not learn. The contour feathers of the last survivor expanded so tardily that when last seen at the age of sixteen days it was still nearly naked, although most nestling Graycaps are fully clothed when fourteen days old. A kindhearted girl told me that with tweezers she had extracted eight *tórsalos* from a *pecho amarillo* (yellow-breast), who may have been either a Gray-capped or a Vermilion-crowned flycatcher, and that the nestling survived the treatment. Usually the larvae emerge and the swellings subside before the young birds fly.

Lice, of which hundreds of species infest birds, are sometimes troublesome, especially to Eastern Phoebes, who occupy the same nest in a sheltered spot year after year, with annual increments. Phoebes and Barn Swallows also use each other's unoccupied nests as foundations for their own, occasionally building upward to form a compound structure with the two kinds of nests alternately superimposed. In *Birds of an Iowa Dooryard,* Althea Sherman told of her struggles to rid of lice the nest of a succession of phoebes that for many years resided in her barn. I have found a heavy infestation of lice only in a gourd occupied for several broods by a Southern House Wren. An advantage of building a new nest for each brood, or at least for each breeding season, amid vegetation rather than in a hole or nook—as most flycatchers do—is relative freedom from lice and other small nuisances.

Of larger parasites, we have already met, in chapter 7, the Piratic Flycatcher, which steals covered nests of other birds. To the victim of nest piracy, this may bring repeated disaster; for if the Pirates that have taken her first nest lose eggs or young from it, they are likely to seize the replacement nest that she has built nearby.

More widespread and numerous are cowbirds, which by depositing their eggs into the nests of other birds induce them to rear the parasites' young. Eggs of the Brown-headed Cowbird have been found in nests of twenty-two of the thirty-four species of flycatchers that breed in the United States and Canada. Most of the remaining twelve breed in holes or enclosed nests, or have very limited ranges north of Mexico, so that few of their nests have been examined. Fourteen of the twenty-two species in which cowbirds' eggs have been found, of which six were small species of *Empidonax*, are known to have reared the cowbirds. Even the Eastern Kingbird, which commonly ejects cowbird eggs from its nest, occasionally overlooks them and nurtures the alien young. Unlike hatchling cuckoos of a number of species, cowbirds do not heave out of the invaded nest every egg or nestling that they find there, but often grow up with the host's progeny. However, the latter are greatly disadvantaged, especially when they are flycatchers, as flycatchers' eggs take several days longer to hatch than do cowbirds' eggs. Born first, the cowbird is much bigger than the hatchling flycatcher and can intercept most of the food that the parents bring, to the great detriment of their own nestlings, which often succumb.

One of the most heavily parasitized flycatchers is the Eastern Phoebe. In some regions a third of its nests receive cowbirds' eggs. Years ago, I followed events in a phoebe's nest with an intruded egg. When found beneath a bridge on May 20, the nest contained one newly hatched cowbird and four phoebe eggs, two of which were barely pipped. By May 22, when two of the phoebe's nestlings had hatched, the cowbird, several times their size, had opening eyes and sprouting wing feathers. By May 25, one phoebe nestling and one of the unhatched eggs had vanished. On May 28 the young cowbird, already well feathered, could perch and showed fear; the surviving phoebe nestling, still blind and with pinfeathers just sprouting, was weak and helpless. The next day the cowbird was perching on the nest's rim; the phoebe's eyes were just opening. By May 30 the cowbird had left the nest; the phoebe's eyes were half open. It did not leave the nest until June 5 or 6, at which stage the parents were still feeding the nestling cowbird. The outcome of this nest was typical. In a study in Michigan, only 9 percent of 323 phoebes' eggs in parasitized nests produced fledglings, as opposed to 46 percent of 1,203 phoebes' eggs in unparasitized nests. Of 139 cowbirds' eggs in these phoebes' nests, 41.7 percent yielded fledged young.

In the southwestern United States, Mexico, and Central America, Bronzed Cowbirds have foisted their eggs into the nests of six species of flycatchers, including the redoubtable Great Kiskadee and the vigilant Tropical Kingbird. None of the six is known to have raised Bronzed Cowbirds, but this may be because of paucity of observers. In South America, Shiny Cowbirds have intruded their eggs into nests of thirty-two species of flycatchers, a number which would certainly increase if the birds of this continent so rich in flycatchers were more thoroughly investigated. Ten of the parasitized species, including the Great Kiskadee, Fork-tailed Flycatcher, Tropical Kingbird, and Cattle Flycatcher, are known to have raised Shiny Cowbirds.

In the localities where I have chiefly studied flycatchers, cowbirds were absent, but many other enemies preyed upon eggs and nestlings. The first seven lines in table 4 give the outcome of 207 open or roofed (not hanging) nests of seven species in the valley of El General on the Pacific slope of southern Costa Rica, between two thousand and three thousand feet above sea level. These nests were in gardens, coffee plantations, pastures, weedy fields, and light second-growth woods near but not in the shrinking rain forest,

TABLE 4   *Nesting Success of Flycatchers with Open or Roofed Nests*

Species	Nests found at all stages			Eggs in nests found early			
	Number	Produced young	Percent successful %	Laid	Hatched	Young fledged	Egg Success %
Costa Rica							
Mistletoe Flycatcher	39	16	41	30	14	12	40
Yellow-bellied Elaenia	36	15	42	36	13	6	17
Lesser Elaenia	39	11	28	37	13	8	22
Boat-billed Flycatcher	19	8	42	—	—	—	—
Vermilion-crowned Flycatcher	16	4	25	29	12	6	21
Gray-capped Flycatcher	40	18	45	76	47	31	41
Tropical Kingbird	18	5	28	18	8	1	6
7 species	207	77	37	226	107	64	28
Ecuador							
Tawny-crowned Pygmy-Flycatcher	22	13	59	38	26	19	50
Vermilion Flycatcher	188	93	49	330	263	151	46
Short-tailed Field Flycatcher	36	14	39	103	59	41	40
Snowy-throated Kingbird	29	21	73	70	—	51	73
4 species	275	141	51	541	348	262	48
Temperate North America							
Acadian Flycatcher	66	39	59	168	101	90	54
Willow Flycatcher[a]	92	60	65	302	223	198	66
Willow Flycatcher[b]	91	36	40	272	149	99	36
Least Flycatcher	16	9	56	56	42	34	61
Eastern Phoebe[a]	194	126	65	867	621	531	61
Eastern Phoebe[b]	110	88	80	502	392	348	69
Say's Phoebe	45	30	67	169	124	96	57
5 species	614	388	63	2336	1652	1396	60

Sources and localities: Costa Rica—Skutch 1966; Ecuador—Marchant 1960; temperate North America— Walkinshaw 1961 (Acadian and Least flycatchers, Michigan); Walkinshaw 1966 (Willow Flycatcher[a], Michigan); Holcomb 1972 (Willow Flycatcher[b], USA) Weeks 1979 (Eastern Phoebe[a], Indiana), Faanes 1980 (Eastern Phoebe[b], Wisconsin); Ohlendorf 1976 (Say's Phoebe, Texas).

which not long ago covered nearly all of this region. The first three columns pertain to nests of known outcome at whatever stage they were found. Nests discovered after incubation had begun are already a selected group, having escaped perils to which other nests have succumbed. The next three columns give records for those of the nests in the columns on the left that were found before incubation began and thus, for our purposes, were exposed to losses for longer intervals. Only 28 percent of the eggs in these nests yielded fledglings, whereas 37 percent of the nests found at all stages produced at least one fledged young. This appears to be very low nest success, yet it differs little from that of a sample of twenty-three species, from doves to sparrows and including the same flycatchers, in the same localities during the same years. The egg success of the twenty-three species was 30 percent, only 2 percent higher than that of the seven flycatchers considered alone, and it differs little from that in other parts of the humid tropics.

The records for the next four flycatchers in table 4 were made by S. Marchant on the Santa Elena Peninsula of southwestern Ecuador, an arid region of erratic rainfall. Some of the nests on the right side of the table were found at various stages of incubation but all before the eggs hatched. Accordingly, these Ecuadorian records are not strictly comparable with those from Costa Rica, but the difference is probably slight. That birds in the arid region enjoy substantially higher success than those in rain-forested regions is also indicated by the outcome of nests found at all stages. The 51 percent success is close to the average for passerine birds in the North Temperate Zone. Amid the lush vegetation of rainy regions, birds have much better opportunities to hide their nests than in the sparse growth of arid lands, but predators appear to be more numerous.

The last seven lines in table 4 are based upon studies made in temperate North America by several naturalists in the localities indicated in the footnote. These northern flycatchers, with more accelerated nesting schedules, lost fewer nests and eggs than did those of either the humid or the arid tropics.

The relatively few flycatchers' nests that one finds in rain forests are mostly inaccessible to examination. The flycatcher with open nests inside the rain forest for which I have most records is the Golden-crowned Spadebill, for which I know the outcome of eleven nests found at all stages. Five failed and six (55 percent) produced fledglings, which is a very good record for birds in this predator-ridden environment.

One might suppose that nests dangling in a clear space where they are difficult for all but aerial predators to reach would be more successful than cuplike nests. Greater safety should compensate for more laborious and time-consuming building. Unfortunately, it is difficult to learn the outcome of significant numbers of nests no more accessible to inquiring bipeds than to flightless quadrupeds. I know the outcome of eighteen of the blackish, retort-shaped nests of the Yellow-olive Flycatcher that hang conspicuously at the forest's edge or from scattered trees beyond it; eight, or 44 percent, yielded flying young. The thin, tough strands of which these nests are composed are so attractive to building birds of other species that many nests are pulled apart before they are finished; these prematurely destroyed nests are not included in calculations of success. Of ten Sulphur-rumped Flycatchers' pensile nests in the forest or beside it, five produced fledglings. Of six Royal Flycatchers' greatly elongated hanging nests, four or possibly five were successful. These few records suggest that dangling nests are safer than others, except those in holes. However, of eleven Ochre-bellied Flycatchers' beautiful mossy nests in Costa Rica, only four, or 36 percent, escaped disaster, and in Trinidad only four of thirty-three nests, or 12 percent, produced fledglings. The subject merits further study.

The poor success of birds' nests in the humid tropics is compensated by the greater longevity of adults, which neither undertake hazardous migrations nor confront the rigors of snowy winters. On the island of Trinidad, three Ochre-bellied Flycatchers banded by David and Barbara Snow survived at least for six, 8.5, and 8.5 years. A tiny White-throated Spadebill was still alive nine years after it was banded, and a slightly bigger Forest Elaenia after ten years. In the United States, some longevity records determined by banding and recapture are: four years for the Yellow-bellied, Willow, and Least flycatchers, with another record of five years for a Least Flycatcher; seven years for Eastern and Western kingbirds, the Eastern Wood-Peewee, and Alder Flycatcher; eight and nine years for Eastern Phoebes; and eleven years for a Great Crested Flycatcher. These records are minima, for the birds were still alive when released after recapture. The potential life span of birds, rarely realized in the wild where they are beset by many perils, is often demonstrated in well-conducted aviaries; but flycatching birds are so difficult to keep in confinement that how long they might live in optimum conditions has apparently not been learned.

11

Flycatchers as Neighbors

The valiant, active bird that we now call the Eastern King-bird was probably one of the first members of the flycatcher family to be noticed by English colonists in North America. In *The Natural History of Carolina, Florida and the Bahama Islands* (1731–43), Mark Catesby called this bird the "Tyrant," a designation that apparently influenced Linnaeus, the founder of our present system of scientific nomenclature, to give this flycatcher the binomial *Lanius Tyrannus,* which might be translated as "shrike tyrant." Since shrikes are notorious for their predatory habits, this combination of names seems to intensify the notion that the Eastern Kingbird is a relentless persecutor of other birds.

Recognizing Linnaeus's error in classifying a flycatcher in a genus of unrelated shrikes, the French naturalist Bernard Germain Etienne de la Ville, Comte de Lacépède, established the genus *Tyrannus* in 1799; our bird became *Tyrannus*

Eastern Kingbird, *Tyrannus tyrannus.*
Sexes alike. Southern Canada and the
United States except the southwest,
winters in South America from
Colombia to northwestern Argentina.

tyrannus—the tyrannical tyrant—and the family to which it belongs became the Tyrannidae. No matter how inappropriate the Latin name of an organism may be, if it has priority, the international rules of scientific nomenclature veto its change, with the result that we have the adjectives *canadensis, mexicanus,* or *chilensis* firmly attached to birds unknown in Canada, Mexico, or Chile—and *tyrannus* applied to a flycatcher that hardly deserves this opprobrious epithet. The English name *kingbird* appears not to be an inaccurate translation of *Tyrannus* but to refer to the bird's usually concealed red crown patch. The false impression of the character of this family conveyed by the name *tyrant flycatchers* is intensified by having their behavior described as "pugnacious," "furious", "vicious," "savage," and "violent" in two of our richest sources of information on their habits, Arthur Cleveland Bent's compilation of the literature on flycatchers of temperate North America and W. H. Hudson's accounts, in *Birds of La Plata,* of those he knew as a boy and young man in Argentina. How just, or widely applicable, are these characterizations of a family of birds most numerous in the tropics, where neither of these authors knew them?

Consider the ways of the Least Flycatcher. When courting in spring, males are, as Bent wrote, "the most active, noisiest, and most pugnacious of any of our small birds." A cause of the frequent conflicts among these little flycatchers is crowding. As among many flycatchers and other birds, each male claims a plot of ground, his territory, to which he tries to attract a mate, and from which he tries to exclude all other males of his own species, fighting if necessary. In favorable areas, Least Flycatchers' territories are small and crowded. Thirty-three of their territories mapped by Peggy MacQueen in Michigan ranged from 0.03 to 0.50 acres. In different years, she found them breeding at a density of 2.7 and 2.0 pairs per acre. In the mountains of Virginia, David E. Davis found eleven pairs in nineteen acres, or a density of 1.7 acres per pair. Crowded male flycatchers compete for mates and frequently invade their neighbors' territories, giving rise to lively pursuits which end in tumbling fights if the trespasser does not promptly return to his own domain. A female does not pursue a solitary intruder unless he advances to within about twenty feet of her nest; but if two birds invade the territory together, she helps her partner to chase them out. Nearly always the invader of an established territory is evicted by these pursuits or fights. After following the defeated bird a few feet beyond the territory, the victor returns to his favorite perch and calls *che-bec*, flicking his tail and raising his crest with each repetition. With larger territories, related Acadian and Willow flycatchers are less often involved in struggles.

In 1936 and 1937, Lesser Elaenias were abundant in a weedy field beside my thatched cabin in the valley of the Río Buena Vista in southern Costa Rica. Three occupied nests were in view from my window, in a triangle with a base of seventy feet and equal sides of about ninety feet. With such mercurial birds so close together, pursuits were frequent. As one elaenia closely pursued another, a third or even a fourth might join the chase. One April morning, while I swept out my cabin with the door ajar, an elaenia trying to escape from a rival fled through the narrow opening and fell almost at my feet, back down, in a corner of the room, with its assailant on top. In an instant, before either had suffered any evident injury, they arose and went their separate ways—the usual outcome of these affrays. The onslaught was violent, but the birds' fury was soon spent. Although I often saw two clutch together and tumble about on the ground, they always separated after a few seconds, apparently none the worse for their tussles. These encounters prompted me to use the name "Bellicose Elaenia" for this species in the second volume of *Life Histories of Central American Birds.*

In pastures with scattered trees in the Costa Rican highlands, little, dull Mountain Elaenias were abundant and quarrelsome. Their frequent pursuits, in which two, three, or even four birds participated, often ended on the pasture grass. While one lay there, its assailant repeatedly dashed down on it from about a foot above. On another occasion, four elaenias fell in a cluster from a tree to the ground, where two lay close together while a third pounced upon one or both of them from a height of a few inches, repeating this several times. Meanwhile, the fourth remained upon the grass near the victims of the assault. As usual, in less than a minute all flew away with no sign of injury. It is significant that like the Least Flycatcher, both of these belligerent elaenias are migratory or at least wandering birds, which claim territories and acquire mates as the breeding season approaches. The Mountain Elaenia travels up and down the mountains with the changing seasons; the Lesser Elaenia appears to go farther, but its movements have not been traced. The sedentary, constantly mated Yellow-bellied Elaenia impresses me as a far milder, more pacific bird. I have never seen this species, nor indeed any of the permanently resident flycatchers amid which I have long dwelt, engage in such conflicts as are frequent among the two migratory elaenias. With more time to arrange their affairs, they settle things with less tumult.

Defense of territory and protection of the nest are quite different matters. The latter is practiced, to the best of a bird's ability, against all creatures, feathered, furry, or scaly, that approach the sacred precincts too closely; the former, as a rule, only against other individuals of the same species. Interspecific territorialism, involving two species, is much rarer than intraspecific territorialism, involving only one. In interspecific territorialism, members of a second species are treated, as regards spatial relations, like other individuals of the same species; often they are closely related, resemble each other in plumage, notes, or displays, or compete for the same resources. Among flycatchers, interspecific territorialism has been reported of two sibling species of *Empidonax*, the Gray and the Dusky flycatchers, which look much alike. As mapped in California by Ned Johnson, the mutually exclusive territories of these two little birds remind one of a mosaic composed of rounded tiles of different colors.

More curious and complex were the relations between Least Flycatchers and American Redstarts, as studied in New Hampshire by Thomas Sherry and Richard Holmes. When, at the age of two years, the male redstart acquires his full breeding attire of black and orange, this wood warbler is a far more brilliant bird than the little olive flycatcher, but both flit through the trees catching small insects. The slightly heavier flycatchers are dominant over the redstarts and by aggressive behavior tend to exclude these competitors from their territories. Yearling redstarts, which breed in plumage intermediate between that of older males and the much duller females, manage better than their elders to coexist with the flycatchers, with the result that territories of the two species more frequently overlap.

The plainly clad *Empidonax* flycatchers, widespread in temperate North America and numerous in both species and individuals, have received more attention by ornithologists, professional and amateur, than have members of any other flycatcher genus, and assessments of their temperaments have differed widely. After an intensive study of Willow Flycatchers in southeastern Washington, James King wrote: "The reputed pugnacity of Traill's [Willow] Flycatchers toward birds of other species . . . is either much overrated or does not apply to the species in this area. I have recorded only one instance of such behavior, the harassing of an Eastern Kingbird."

Perhaps this kingbird deserved such treatment, for some individuals, like the male of a pair that nested in a dead tree rising above a pond in New Hampshire, where they were watched by Lawrence Kilham, are overzealous in the defense of their nesting area. Keeping vigilant watch over a wide expanse of open water and surrounding shores, this kingbird attacked birds no more harmful than swallows and goldfinches that flew at a distance from his nest. Other Eastern Kingbirds are less aggressive. One of Bent's correspondents wrote of a pair of kingbirds nesting in Mississippi in a large pecan tree, which also held nests of the Eastern Wood-Pewee, Baltimore Oriole, Red-eyed Vireo, and two pairs of House Sparrows. All these birds appeared to live in more or less perfect accord, except the wood-pewee, its nest too near that of one pair of sparrows.

Such tolerance of nesting associates is widespread among flycatchers, which do not repel birds of other kinds that build nests near their own, gaining protection from predators by proximity to birds that so vigorously repulse them. In southern Texas, four of ten woven pouches of Altamira Orioles were built within three or four yards of occupied Great Kiskadees' nests, and one was also close beside a Tropical Kingbird's nest. The latter chased away Great-tailed Grackles, which prey on eggs and nestlings. The kiskadees drove off parasitic Bronzed Cowbirds and other birds, and appeared to give the orioles confidence to enter their nests in the presence of Barbara Pleasants, who reported this nesting association.

In Arizona, tanagers, wood warblers, vireos, and other birds attend their eggs and young close to nests of vigilant Greater Pewees. Elsewhere, a Least Flycatcher built her nest about five feet from an incubating wood-pewee. Nests of a Mourning Dove, Orchard Oriole, House Sparrow, and three different Northern Mockingbirds were found in the same trees with those of Scissor-tailed Flycatchers. Western Kingbirds lived "in perfect harmony" with Bullock's Orioles, Mourning Doves, Yellow Warblers, and House Sparrows. A nesting House Finch shared a mesquite tree with a Vermilion Flycatcher. More surprisingly, Scissor-tailed Flycatchers and Red-tailed Hawks nested on different sides of a large live oak tree. In Oregon, Major Charles F. Bendire, an early biographer of North American birds, found a pair of Western Kingbirds nesting in a middle-sized pine tree with Swainson's Hawks and Bullock's Orioles, all apparently on excellent terms. He also saw two pairs of Western Kingbirds raising their broods in the same tree, apparently in perfect harmony.

In the tropics, as farther north, flycatchers share their nest trees with a diversity of birds, benefiting their neighbors by their protective vigilance. In southern Tamaulipas, Mexico, George M. Sutton and O. S. Pettingill, Jr., watched twelve nests of Vermilion-crowned Flycatchers, each of which was close to the nest of some other species: six near Rose-throated Becards' nests, four close to Altamira Orioles' nests, one near a Boat-billed Flycatcher's nest, and one near that of a Great Kiskadee. These discoveries led them to call *Myiozetetes similis* the Social Flycatcher, a name that has been accepted by the American Ornithologists' Union. But "social" implies close association by members of the same species, as in the Sociable Weaver and the Gray-capped Social Weaver of Africa. Although Vermilion-crowned Flycatchers and closely related Gray-capped or Rusty-margined flycatchers sometimes nest in the same small tree, two pairs of Vermilion-crowns are not known to share a tree. They are tolerant of other species but no more social than any of the birds that consort with them.

In the valley of the Río Orinoco, George Cherrie found Vermilion-crowned Flycatchers, Great Kiskadees, Piratic Flycatchers, and Yellow Orioles all nesting in a single tree. Nests of the oriole and the kiskadee were frequently close together in a tree. Tropical Kingbirds and Blue-gray Tanagers nested only two yards apart. In Colombia, José Borrero found nests of the Pied Marsh-Flycatcher, Great Kiskadee, Tropical Kingbird, and a species of *Myiozetetes* all in the same mango tree.

In a small orange tree on Barro Colorado Island in Gatún Lake, Panama, a Tropical Kingbird, a Vermilion-crowned Flycatcher, and a Bananaquit built their nests. The kingbirds did not molest the much smaller Bananaquit, but when the male kingbird, watching from a dead twig at the top of the tree, saw the Vermilion-crown approaching with material for her nest, he often darted toward her. The smaller flycatcher usually turned tail and fled unresistingly, although at times she and her mate showed their spirit by displaying their vermilion crown patches. Frequently the female Vermilion-crown would dart past the kingbird and take her billful of material into her nest, which he never disturbed. Thus, for all the kingbird's bluster, the Vermilion-crown finished her nest close below his own.

Near the end of a long, narrow cove in the wooded shoreline of the same island, a decaying trunk, about twenty-five feet high, rose above the still water about one hundred feet from the shore. Attached to the rotting wood grew a great variety of epiphytes, including aroids, orchids, ferns, and a small, wide-spreading bush. From the branches of this shrub hung the long, woven pouches of a small colony of Yellow-rumped Caciques. Just below these swinging nests a pair of Tropical Kingbirds were feeding nestlings in a shallow open cup tucked amid the epiphytes. Nearer the water a pair of Rusty-margined Flycatchers had built a bulky roofed nest. The old trunk also supported a diversity of nests of wasps and little, black, stingless meliponine bees. Although the kingbirds occasionally displaced male caciques from the coveted lookout at the very top of the bush—much as the

caciques supplanted each other—all these so different creatures dwelt together peacefully enough until one morning
when, for reasons unknown, the bees became highly excited.

Gyrating and circling about each other, thousands of
bees formed a dark, troubled cloud hanging stationary in
the air in front of the nests. Soon the bees began to attack the
nestling kingbirds, which fortunately already had enough
plumage to give some protection from the bees' biting
mandibles. Frantically flapping their wings, hopping from
side to side of their narrow nest, the nestlings tried to pick
off the bees with their bills. Powerless against so multitudi-
nous a swarm, the parents could only look on from a bar-
rigon tree on the neighboring shore, vibrating uplifted wings
and twittering to each other, as they usually did when ex-
ited. As though to relieve their feelings, the parents darted
with clacking bills at the caciques, especially the bigger
males, and made them flee—much as I have seen other
birds, enraged by an apparent threat to their nest by a creature they hesitated to attack, vent their anger by tearing
harmless leaves. After continuing for over an hour their un-
provoked attacks on the nestlings, the bees finally quieted
down and withdrew. Peace returned to the community; the
parent kingbirds resumed feeding their nestlings, which a
few days later flew to the shore.

Flycatchers' hostility to predators may benefit other
birds that live or nest at a distance from the flycatchers' nest
trees. In chapter 10 I told how a pair of Great Kiskadees repulsed a snake trying to reach a colony of Montezuma
Oropendolas. As a Swallow-tailed Kite swooped down to
carry off nest and nestlings of a pair of Golden-masked Tanagers, a Tropical Kingbird, who had no nest nearby, pursued
the kite closely and almost prevented the tragedy. One rainy
afternoon in November, while I watched three Boat-billed
Flycatchers, Vermilion-crowns, and other birds catching insects among the trees in my garden, a hawk suddenly
darted down and seized one of the flycatchers. The raptor,
whom I did not see well enough for identification, dropped

with its victim to the ground on the far side of a hedge, where the Boat-bills assailed it with snapping bills while smaller birds fled. This defensive action was the more remarkable because at this season none of these birds was nesting, and the victim was not one of the Boat-bills.

Not only do flycatchers repulse animals that menace their neighbors or their neighbors' nests, they also occasionally help birds of different species to rear their young. An Eastern Phoebe, as her first brood was becoming independent, fed nestling Tree Swallows for about a week, in the face of opposition by the parent swallows. When an Eastern Kingbird fed nestling Baltimore Orioles, the parents at first tried to drive away their uninvited assistant, but soon they became reconciled to its presence. The kingbird continued to attend the young orioles after they left the nest. A Scissor-tailed Flycatcher fed Common Grackles in a neighboring nest; a Least Flycatcher helped to feed Chipping Sparrows in a nest near its own; and an Eastern Wood-Pewee nourished orphaned nestling Eastern Kingbirds.

Flycatchers not only help parent birds of different species but also receive help from them, which would not occur if they were as aggressive toward other birds as they are sometimes reputed to be. A female House Sparrow fed five nestling Western Kingbirds that were also attended by both of their parents. The latter did not oppose the helper's visits to their nest; but the sparrow, becoming increasingly possessive, twice attacked the parent kingbirds as they approached. In Kansas, a female House Sparrow continued for at least a week to bring bread and other food to three fledgling Eastern Kingbirds whose parents had disappeared. This activity was not without hazard to the sparrow, her head was often caught by the closing of the wide mouths into which she thrust her offerings, and she had to struggle to release herself. An almost identical episode was reported from Louisiana where, also, no adult kingbird was seen to visit the three fledglings whom the sparrow fed for at least ten days.

To reach a fair appraisal of the relations of flycatchers

with neighbors of their own and other species, let us glance briefly at some of the counts against them. The Great Shrike-Flycatcher of southern South America—one of the biggest members of the family—and the far more widespread Great Kiskadee prey not only upon lizards, frogs, and small mammals but at least occasionally upon birds' eggs and nestlings. A Great Kiskadee may carry off the pillaged nest of a small bird to increase the bulk of its own untidy structure. Considering the number of weaker birds that breed, apparently often successfully, in the same tree with a kiskadee, this bird can hardly be a frequent plunderer of their nests. Gray Kingbirds and Brown-crested Flycatchers have been seen to catch hummingbirds and knock them against a perch before swallowing them, much as they treat large insects, for which, perhaps, they mistake these tiny victims.

A pair of Sulphur-bellied Flycatchers, their nest having been plundered by jays, fought for a week with a pair of red-shafted Northern Flickers and finally captured their hole with three fresh eggs, above which the female flycatcher built her nest and laid her own. Cattle Flycatchers struggle with other birds for possession of Firewood-gatherers' castles of interlaced sticks, which all need for housing their broods. Cassin's Kingbirds chased blackbirds until they dropped worms gathered from a freshly plowed field, then promptly swallowed their booty. A Scissor-tailed Flycatcher pursued a Lark Sparrow who was trying to catch a flying insect and snatched the creature from beneath the sparrow's bill. All these misdemeanors, if such they may be called, are much more frequent among certain other birds than among flycatchers. Among the tropical flycatchers amid which I have lived for many years, the only one of the foregoing misdeeds that I have witnessed was that of a Great Kiskadee, on a single occasion. Nearly all these tropical flycatchers are well-behaved birds, aggressive only toward creatures that might harm their nests or young.

Often birds attack or mob other birds for no other reason than that they are unfamiliar and possibly harmful.

When a Fork-tailed Flycatcher, a rare vagrant on the Falkland Islands, alighted on a fence near resident Dark-faced Ground-Flycatchers, they harassed its every move by swooping over it without ever striking it. The next day, apparently satisfied that the stranger was innocuous, they left it in peace. O. S. Pettingill, Jr., also told how a ground-flycatcher persisted in following another unfamiliar wanderer from the continent, a Fire-eyed Diucon, until the vagrant turned around and chased its persecutor.

A color pattern, vocalization, or movement of a bird of a different species may cause a territory holder to react as if the visitor were an intruder of its own species. In Arizona, Ash-throated Flycatchers repeatedly attacked and knocked out of the air ground-foraging Cassin's Sparrows flying upward from treetops while they sang. George T. Austin and Stephen M. Russell, who observed this unexpected assault, attributed it to the similarity of the upward-arching curve of the sparrows' song flight to the trajectory of a flycatcher hawking insects, which caused the Ash-throats to treat the sparrows as competitive invaders of their territories.

When not dwelling in wild forests or remote savannas, flycatchers are nearly always good neighbors of humans, rarely clashing with their economic interests. Most prefer wild to cultivated fruits. They appear never to damage field or garden crops but increase productivity by devouring great numbers of injurious insects. In past times, kingbirds, accused of taking too many honeybees, were shot by bee-keepers; but careful studies proved that the bees they ate in small numbers were mostly unproductive drones. Kingbirds now enjoy the legal protection given to other migratory passerine birds, which, unfortunately, does not save them from television towers, against which many flycatchers collide fatally, along with countless thousands of other birds that migrate by night.

Many country dwellers welcome phoebes that nest trustfully on their buildings and roost under their eaves, or kingbirds that attack hawks which might prey upon domes-

Sirystes, *Sirystes sibilator.*
Sexes alike. Eastern Panama
through South America to
northeastern Argentina.

tic chickens. With many similar species that can be distinguished only by careful attention to plumage, habits, and vocalizations, flycatchers offer to birdwatchers opportunities to sharpen their skill in field identification. Even in the tropics where they abound, flycatchers contribute little to the brilliant colors of the feathered world, but to the dedicated student they present a fascinating diversity of lifestyles and nests.

Adaptable and vigorous, flycatchers have spread much more widely over the Western Hemisphere than any of the other six New World families of nonoscine passerines. To know them as American flycatchers would adequately distinguish them from the Old World flycatcher family, without the undeserved, opprobrious designation "tyrant." The bold aggressiveness that earned for the largest avian family in the western hemisphere this derogatory name is exhibited by only a few of its bigger members and is directed mainly against birds and other animals that menace their nests and young, or occasionally against unfamiliar and possibly harmful birds. Rarely, probably in an exuberantly playful mood, flycatchers pursue small, innocuous birds, which they do not injure. Like most birds, they tend to repulse all intruders, big or little, from the immediate vicinity of their nests, but often they are surprisingly tolerant of small and harmless visitors to the nest tree. Birds of other species nest close to flycatchers, protected by their vigilance and valor. Far from being tyrants of an avian community, many of them are its guardians.

Bibliography

Chapter 1. The Flycatcher Family

(General references and works on classification, anatomy, plumage, and migration. Much of the information in the other chapters is also from these works, which are not again cited.)

Bent, A. C. 1942. *Life Histories of North American Flycatchers, Larks, Swallows, and Their Allies*. U.S. National Museum Bulletin No. 179. (Habits and distribution of flycatchers of the United States and Canada.)

Conover, M. R., and D. E. Miller. 1980. Rictal bristle function in Willow Flycatcher. *Condor* 82:469–71.

French, R. 1973. *A Guide to the Birds of Trinidad and Tobago*. Wynnewood, Pa.: Livingston Publishing Company. (Much information on habits, food, and breeding.)

Fitzpatrick, J. W. 1985. "Form, foraging behavior, and adaptive radiation in the Tyrannidae." In *Neotropical Ornithology*, ed. P.A. Buckley, M. S. Foster, E. S. Morton, R. S. Ridgely, and

F. G. Buckley. American Ornithologists' Union Ornithological Monograph no. 36.

Goodall, J. D., A. W. Johnson, and R. A. Phillipi B. 1967. *Las Aves de Chile: Su conocimiento y Sus Costumbres.* Buenos Aires: Platt Establecimientos Gráficos.

Haverschmidt, F. 1968. *Birds of Surinam.* Wynnewood, Pa. Livingston Publishing Company. (Notes on food and breeding.)

Hilty, S. L., and W. L. Brown. 1986. *The Birds of Colombia.* Princeton, N.J.: Princeton University Press. (Habits and breeding.)

Hudson, W. H. 1920. *Birds of La Plata.* 2 vols. London: J. M. Dent and Sons. (Vivid accounts of behavior and nesting of flycatchers in Argentina.)

Johnson, N. K. 1963. Biosystematics of sibling species in the *Empidonax hammondii-oberholseri-wrightii* complex. *University of California Publications in Zoology* 66:79–238.

Lederer, R. J. 1972. The role of avian rictal bristles. *Wilson Bulletin* 84:193–97.

Meyer de Schauensee, R. 1970. *A Guide to the Birds of South America.* Wynnewood, Pa.: Livingston Publishing Company.

Monroe, B. L., Jr. 1968. *A Distributional Survey of the Birds of Honduras.* American Ornithologists' Union Ornithological Monograph no. 7.

Morton, E. S. 1971. Food and migration habits of the Eastern Kingbird in Panama. *Auk* 88:925–26.

Rappole, J. H., and D. W. Warner. 1980. "Ecological aspects of migrant bird behavior in Veracruz, Mexico." In *Migrant Birds in the Neotropics: Ecology, Behavior, Distribution, and Conservation,* ed. A. Keast and E. S. Morton. Washington, D.C.: Smithsonian Institution Press.

Riggs, C. D. 1955. Night migration of the Scissor-tailed Flycatcher. *Condor* 57:310.

Sick, H. 1984. *Ornitologia Brasileira; Uma Introdução.* 2 vols., 2nd. ed. Brasilía: Editora Universidad de Brasília. (Notes on habits and nesting.)

———. 1993. *Birds in Brazil: A Natural History.* Translated by W. Belton. Princeton, N.J.: Princeton University Press.

Skutch, A. F. 1951. Life history of the Boat-billed Flycatcher. *Auk* 68:30–49.

———. 1954. Life history of the Tropical Kingbird. *Proceedings of the Linnaean Society of New York* 63–65:21–38.

————. 1960. *Life Histories of Central American Birds II.* Pacific Coast Avifauna no. 34. (Includes 31 species of flycatchers.)

————. 1967. *Life Histories of Central American Highland Birds.* Nuttall Ornithological Club Publication no. 7. (Includes six species of flycatchers.)

————. 1972. *Studies of Tropical American Birds.* Nuttall Ornithological Club Publication no. 10. (Includes four species of flycatchers.)

————. 1981. *New Studies of Tropical American Birds.* Nuttall Ornithological Club Publication no. 19. (Includes five species of flycatchers.)

Stettenheim, P. 1974. The bristles of birds. *Living Bird* 12: 201–34.

Stiles, F. G., A. F. Skutch, and D. Gardner. 1989. *A Guide to the Birds of Costa Rica.* Ithaca, N.Y.: Cornell University Press. (Much information on habits, food, and breeding.)

Traylor, M. A., Jr., and J. W. Fitzpatrick, 1982. A survey of the Tyrant Flycatchers. *Living Bird* 19:7–50.

Chapter 2. Food and Foraging

Baker, B. W. 1980. Commensal foraging of Scissor-tailed Flycatchers with Rio Grande Turkeys. *Wilson Bulletin* 92:248.

Binford, L. C. 1957. Eastern Phoebes fishing. *Auk* 74:264–65.

Fitzpatrick, J. W. 1980. Foraging behavior of Neotropical tyrant flycatchers. *Condor* 82:43–57.

Greenberg, R., and J. Gradwohl. 1980. Leaf surface specializations of birds and arthropods in a Panamanian forest. *Oecologia* (Berlin) 46:115–24.

Hespenheide, H. A. 1964. Competition in the genus *Tyrannus*. *Wilson Bulletin* 76:265–81.

————. 1971. Food preference and the extent of overlap in some insectivorous birds, with special reference to the Tyrannidae. *Ibis* 113:59–72.

Lefevbre, L., and D. Spahn. 1987. Gray Kingbird predation on small fish (*Poecilia* sp.) crossing a sandbar. *Wilson Bulletin* 99:291–92.

Murphy, M. T. 1987. The impact of weather on kingbird foraging behavior. *Condor* 89:721–30.

Skutch, A. F. 1980. Arils as food of tropical American birds. *Condor* 82:31–42.

Verbeek, N. A. M. 1975a. Comparative feeding behavior of three coexisting tyrannid flycatchers. *Wilson Bulletin* 87: 231–40.

———. 1975b. Northern wintering of flycatchers and residency of Black Phoebes in California. *Auk* 92:737–49.

Willis, E. O. 1966. The role of migrant birds at swarms of army ants. *Living Bird* 5:187–231.

Chapter 3. Daily Life

Allan, P. F. 1950. Scissor-tailed Flycatcher, *Muscivora forficata*, feeding at night. *Auk* 67:517.

Fitch, F. W., Jr. 1947. The roosting of the Scissor-tailed Flycatcher. *Auk* 64:616.

Munn, C. A. 1985. "Permanent canopy and understory flocks in Amazonia: Species composition and population density." In *Neotropical Ornithology,* ed. P. A. Buckley, M. S. Foster, E. S. Morton, R. S. Ridgely, and F. G. Buckley. American Ornithologists' Union Ornithological Monograph no. 36.

Nice, M. M. 1931. *The Birds of Oklahoma.* Norman: University of Oklahoma Press.

Pearson, O. P. 1953. Use of caves by hummingbirds and other species at high altitudes in Peru. *Condor* 55:17–20.

Sick, H., and D. M. Teixeira. 1981. Nocturnal activities of Brazilian hummingbirds and flycatchers at artificial illumination. *Auk* 98:191–92.

Slessers, M. 1970. Bathing behavior of land birds. *Auk* 87:91–99.

Chapters 4, 5, and 6. Voice, Displays, and Courtship

De Benedictis, P. 1966. The flight song display of two taxa of Vermilion Flycatcher, genus *Pyrocephalus. Condor* 68:306–7.

Kroodsma, D. E. 1984. Songs of the Alder Flycatcher (*Empidonax alnorum*) and Willow Flycatcher (*Empidonax traillii*) are innate. *Auk* 101:13–24.

Lawrence, L. de K. 1989. *To Whom the Wilderness Speaks.* Toronto, Ontario: Natural Heritage-Natural History.

MacQueen, P. M. 1950. Territory and song in the Least Flycatcher. *Wilson Bulletin* 62:194–205.

McCabe, R. A. 1951. The song and song-flight of the Alder Flycatcher. *Wilson Bulletin* 63:89–98.

Smith, W. J. 1967. Displays of the Vermilion Flycatcher (*Pyrocephalus rubinus*). *Condor* 69:601–5.

———. 1970. Courtship and territorial displaying in the Vermilion Flycatcher, *Pyrocephalus rubinus*. *Condor* 72:488–91.

———. 1971. Behavioral characteristics of Serpophagine Tyrannids. *Condor* 73:259–86.

Snow, B. K., and D. W. Snow. 1979. The Ochre-bellied Flycatcher and the evolution of lek behavior. *Condor* 81:286–92.

Weydemeyer, W. 1973. Singing habits of Traill's Flycatcher in northwestern Montana. *Wilson Bulletin* 85:276–82.

Chapters 7, 8, and 9. Reproduction

Alvarez del Toro, M. 1965. The nesting of the Belted Flycatcher. *Condor* 67:339–43.

Berger, A. J. 1966. The nestling period of the Great Crested Flycatcher. *Wilson Bulletin* 78:320.

Blancher, P. J., and R. J. Robertson. 1982. A double-brooded Eastern Kingbird. *Wilson Bulletin* 94:212–13.

Borrero H., J. I. 1972. Historia natural del titiribí, *Pyrocephalus rubinus* (Aves, Tyrannidae) en Colombia, con notas sobre su distribución. *Mittheilungen Instituto Colombo-Alemán de Investigaciones Científicas* 6: 113–33.

———. 1973. Sobre la historia natural de la viudita, *Fluvicola pica* (Boddaert) (Tyrannidae). *Ardeola* 19:69–87.

Briskie, J. V., and S. G. Sealy. 1987. Polygyny and double brooding in the Least Flycatcher. *Wilson Bulletin* 99:492–94.

———. 1989. Determination of clutch size in the Least Flycatcher. *Auk* 106:269–78.

Cade, T. J., and C. M. White. 1973. Breeding of Say's Phoebe in Arctic Alaska. *Condor* 75:360–61.

Carvalho, C. T. de. 1960. Comportamento de *Myiozetetes cayanensis* e notas biológicas sobre espécies afins (Passeres, Tyrannidae). *Papéis Avulsos do Departamento de Zoologia, Secretaria da Agricultura, São Paulo, Brasil* 14: 121–32.

Crouch, J. E. 1959. Vermilion Flycatchers nesting in San Diego County, California. *Condor* 61:57.

Davis, D. E. 1954. The breeding biology of Hammond's Flycatcher. *Auk* 71:164–71.

———. 1959. Observations of territorial behaviour of Least Flycatchers. *Wilson Bulletin* 71:73–85.

Davis, J., G. F. Fisler, and B. S. Davis. 1963. The breeding biology of the Western Flycatcher. *Condor* 65:337–82.

De Kiriline, L. 1948. Least Flycatcher. *Audubon* 50:149–53.

Eckhardt, R. C. 1976: Polygyny in the Western Wood-Pewee. *Condor* 78:561–62.

Fitch, F. W., Jr. 1950. Life history and ecology of the Scissor-tailed Flycatcher, *Muscivora forficata. Auk* 67:145–68.

Goodpasture, K. A. 1953. Wood Pewee builds with green leaves. *Wilson Bulletin* 65:117–18.

Gross, A. O. 1950. Nesting of the Streaked Flycatcher in Panama. *Wilson Bulletin* 62:183–93.

———. 1964. Nesting of the Black-tailed Flycatcher on Barro Colorado Island. *Wilson Bulletin* 76:248–56.

Haverschmidt, F. 1950. The nest and eggs of *Tolmomyias poliocephalus. Wilson Bulletin* 62:214–16.

———. 1954. The nesting of the Ridgway Tyrannulet in Surinam. *Condor* 56:139–41.

———. 1955. Notes on the life history of *Todirostrum maculatum* in Surinam. *Auk* 72:325–31.

———. 1970. Notes on the life history of the Mouse-colored Flycatcher in Surinam. *Condor* 72:374–75.

———. 1971. Notes on the life history of the Rusty-margined Flycatcher in Surinam. *Wilson Bulletin* 83:124–28.

———. 1974. Notes on the life history of the Yellow-breasted Flycatcher in Surinam. *Wilson Bulletin* 86:215–20.

———. 1978. The duration of parental care in the Common Tody-Flycatcher. *Auk* 95:199.

Kendeigh, S. C. 1952. *Parental Care and Its Evolution in Birds.* Illinois Biological Monographs, vol. 22. Urbana: University of Illinois Press.

Ligon, J. B. 1971. Notes on the breeding of the Sulphur-bellied Flycatcher in Arizona. *Condor* 73:250–52.

Maclean, G. I. 1969. The nest and eggs of the Chocolate Tyrant, *Neoxolmis ruficollis* (Vieillot). *Auk* 86:144–45.

Marchant, S. 1960. The breeding of some S. W. Ecuadorian birds. *Ibis* 102:349–82.

McNeil, R., and A. Martínez. 1968. Notes on the nesting of the Short-tailed Pygmy-Tyrant *(Myiornis ecaudatus)* in northeastern Venezuela. *Condor* 70:181–82.

Morehouse, E. L., and R. Brewer. 1968. Feeding of nestling and fledgling Eastern Kingbirds. *Auk* 85:44–54.

Morton, M. L., and M. E. Pereyra. 1985. The regulation of egg temperatures and attentiveness patterns in the Dusky Flycatcher (*Empidonax oberholseri*). *Auk* 102:25–37.

Mumford, R. E. 1962. Notes on Least Flycatcher behavior. *Wilson Bulletin* 74:98–99.

———. 1964. The breeding biology of the Acadian Flycatcher. *Miscellaneous Publications of the Museum of Zoology, University of Michigan* no. 125:1–50.

Murphy, M. T. 1983. Nest success and nesting habits of Eastern Kingbirds and other flycatchers. *Condor* 85:208–19.

———. 1988. Comparative reproductive biology of kingbirds (*Tyrannus* spp.) in eastern Kansas. *Wilson Bulletin* 100:357–76.

Newman, D. L. 1958. A nesting of the Acadian Flycatcher. *Wilson Bulletin* 70:130–44.

Nice, M. M., and N. E. Collias. 1961. A nesting of the Least Flycatcher. *Auk* 78:145–49.

Parker, T. A., III. 1984. Notes on the behavior of *Ramphotrigon* Flycatchers. *Auk* 101:186–88.

Prescott, D. R. C. 1986. Polygyny in the Willow Flycatcher. *Condor* 88:385–86.

Ricklefs, R. E. 1980. "Watch-dog" behaviour observed at the nest of a cooperative breeding bird, the Rufous-margined Flycatcher *Myiozetetes cayanensis*. *Ibis* 122:116–18.

Sedgwick, J. A., and F. L. Knopf. 1989. Regionwide polygyny in Willow Flycatchers. *Condor* 91:473–75.

Sherry, T. W. 1986. Nest, eggs, and reproductive behavior of the Cocos Flycatcher. *Condor* 88:531–32.

Skutch, A. F. 1976. *Parent Birds and Their Young.* Austin: University of Texas Press.

Smith, W. P. 1942. Nesting habits of the Eastern Phoebe. *Auk* 59:410–17.

Stafford, M. D. 1986. Supernumerary adults feeding Willow Flycatcher fledglings. *Wilson Bulletin* 98:311–12.

Taylor, W. K., and H. Hanson. 1970. Observations on the breeding biology of the Vermilion Flycatcher in Arizona. *Wilson Bulletin* 82:315–19.

Thomas, B. T. 1979. Behavior and breeding of the White-bearded Flycatcher (*Conopias inornata*). *Auk* 96:767–75.

Walkinshaw, L. H., and C. J. Henry. 1957. Yellow-bellied Flycatcher nesting in Michigan. *Auk* 74:293–304.

Weeks, H. P., Jr. 1977. Nest reciprocity in Eastern Phoebes and Barn Swallows. *Wilson Bulletin* 89:632–35.

———. 1978. Clutch size variation in the Eastern Phoebe in southern Indiana. *Auk* 95:656–66.

CHAPTER 10. ENEMIES, DEFENSE, NESTING SUCCESS, AND LONGEVITY

Beebe, C. W. 1905. *Two Bird-lovers in Mexico.*

Belt, T. 1888. *The Naturalist in Nicaragua.* 2nd. ed. London: Edward Bumpus.

Faanes, C. A. 1980. Breeding biology of Eastern Phoebes in northern Wisconsin. *Wilson Bulletin* 92:107–10.

Friedmann, H., and L. F. Kiff. 1985. The parasitic cowbirds and their hosts. *Proceedings of the Western Foundation of Vertebrate Zoology* 2:226–302.

Friedmann, H., L. F. Kiff, and S. I. Rothstein. 1977. A further contribution to knowledge of the host relations of the parasitic cowbirds. *Smithsonian Contributions to Zoology* no. 235:1–75.

Holcomb, L. C. 1972. Nest success and age-specific mortality in Traill's Flycatchers. *Auk* 89:837–41.

Janzen, D. H. 1969. Birds and the ant x acacia interaction in Central America, with notes on birds and other myrmecophytes. *Condor* 71:240–56.

Marchant, S. 1960. The breeding of some S. W. Ecuadorian birds. *Ibis* 102:349–82.

Ohlendorf, H. M. 1976. Comparative breeding ecology of phoebes in trans-Pecos Texas. *Wilson Bulletin* 88:255–71.

Sherman, A. R. 1952. *Birds of an Iowa Dooryard.* Boston, Mass.: Cristopher Publishing House.

Skutch, A. F. 1966. A breeding bird census and nesting success in Central America. *Ibis* 108:1–16.

———. 1985. "Clutch size, nesting success, and predation on nests of neotropical birds, reviewed." In *Neotropical Ornithology,* ed. P. A. Buckley, M. S. Foster, E. S. Morton, R. S. Ridgely, and F. G. Buckley. American Ornithologists' Union Ornithological Monograph no. 36.

Snow, D. W., and A. Lill. 1974. Longevity records for some Neotropical land birds. *Condor* 76:262–67.

Terres, J. K. 1980. *The Audubon Society Encyclopedia of North American Birds.* New York: Alfred A. Knopf.

Walkinshaw, L. H. 1961. The effect of parasitism by the Brown-headed Cowbird on *Empidonax* flycatchers in Michigan. *Auk* 78:266–68.

———. 1966. Summer biology of Traill's Flycatcher. *Wilson Bulletin* 78:31–46.

Weeks, H. P., Jr. 1979. Nesting ecology of the Eastern Phoebe in southern Indiana. *Wilson Bulletin* 91:441–54.

Chapter 11. Flycatchers as Neighbors

Austin, G. T., and S. M. Russell. 1972. Interspecific aggression of Ash-throated Flycatchers on Cassin's Sparrows. *Condor* 74:481.

Cherrie, G. K. 1916. A contribution to the ornithology of the Orinoco region. *Museum of the Brooklyn Institute of Arts and Sciences Science Bulletin* 2:133a-374.

Gamboa, G. J. 1977. Predation of Rufous Hummingbird by Wied's Crested Flycatcher. *Auk* 94:157–58.

Gress, B. 1985. House Sparrow found feeding Western Kingbird Nestlings. *Bulletin of the Kansas Ornithological Society* 36:25–26. Abstract in *Journal of Field Ornithology* 58:247. 1987.

Huber, M. R., and J. B. Cope. 1973. House Sparrow dispossesses nesting Eastern Kingbirds. *Wilson Bulletin* 85:338–39.

Kilham, L. 1988. *On Watching Birds.* Chelsea, Vt. Chelsea Green Publishing Company.

King, J. R. 1955. Notes on the life history of Traill's Flycatcher (*Empidonax traillii*) in southeastern Washington. *Auk* 72:148–73.

Nicholson, D. J. 1948. Nest-robbing behavior of the Purple Martin. *Auk* 65:600–1.

Pettingill, O. S., Jr. 1974. Passerine birds of the Falkland Islands: Their behavior and ecology. *Living Bird* 12 (for 1973): 95–136.

Pleasants, B. Y. 1981. Aspects of the breeding biology of a subtropical oriole, *Icterus gularis*. *Wilson Bulletin* 93:531–37.

Sherry, T. W. 1979. Competitive interactions and adaptive strategies of American redstarts and Least Flycatchers in a northern hardwoods forest. *Auk* 96:265–83.

Sherry, T. W., and R. T. Holmes. 1988. Habitat selection by breeding American Redstarts in response to a dominant competitor, the Least Flycatcher. *Auk* 105:350–64.

Skutch, A. F. 1987. *Helpers at Birds' Nests: A Worldwide Survey of Cooperative Breeding and Related Behavior*. Iowa City: University of Iowa Press.

Sutton, G. M., and O. S. Pettingill, Jr. 1942. Birds of the Gómez Farias Region, Southwestern Tamaulipas. *Auk* 59:1–34.

Index

DATE DUE

GAYLORD			PRINTED IN U.S.A